FINITE MATHEMATICS

FINITE MATHEMATICS

JOHN M. PETERSON
Brigham Young University

HOLT, RINEHART AND WINSTON, INC.
New York Chicago San Francisco Atlanta
Dallas Montreal Toronto London Sydney

Library of Congress Cataloging in Publication Data
Peterson, John M, 1938–
Finite Mathematics
1. Mathematics — 1961–   I. Title.
QA39.2.P5   510   73-10457

ISBN 0-03-091298-9
Printed in the United States of America
4 5 6 7 8   0 3 8   1 2 3 4 5 6 7 8 9

PREFACE

This book is an intuitive approach to finite mathematics with introductory algebra as its only prerequisite. It is designed for one three-semester-hour course or one five-quarter-hour course with emphasis on precalculus mathematics that can be used in today's world as opposed to mathematics needed for calculus.

While there is no lack of continuity throughout the book, special effort has been made to make each chapter very nearly "self-contained" to allow for maximum flexibility of use. Thus, chapters may be deleted or covered out of sequence. One exception is Chapter 6, "Vectors and Matrices," which provides the tools necessary for Chapter 7, "Linear Programming and the Theory of Games." Each chapter also contains a summary that lists all of the major concepts, definitions, and formulas introduced in that chapter, as well as a review exercise set. Appendix A is included for those who have an insufficient background in the graphing of inequalities or need such a review before solving linear programming problems. Appendix B is included for those who wish to prove theorems that require mathematical induction.

The objective of this book is to make mathematical concepts understandable, useable, and interesting for the student. The book is, as one reviewer stated, "written for the student."

I express my appreciation to all those who reviewed the manuscript for this book in its developmental stages, and particularly to Michael Sullivan of Chicago State University and Joseph Dorset of St. Petersburg Junior College, St. Petersburg, Florida for detailed reviews and helpful suggestions. I also appreciate the help and advisement of Robert Linsenman and all others of Holt, Rinehart and Winston, Inc. who have assisted with the preparation of this book. Finally, I especially appreciate my wife, Corinne, for her encouragement and typing, and my daughters for their patience in giving up time that was rightfully theirs.

JOHN M. PETERSON

CONTENTS

FINITE MATHEMATICS

chapter

LOGIC

1.1 INTRODUCTION

The word "logic" is common to the vocabulary of almost all people, yet the study of logic is usually found only in courses of either mathematics or philosophy. These two areas of study have a great deal in common due to their logical bases and the great number of historical figures who were prominent in both areas simultaneously. Both fields of study are concerned with more than the superficial question of "what?" and delve into the more pertinent question of "why?"—the answer to which is a logical development of basic assumptions. Thus, logic is used at all levels and in all areas of mathematics, and an understanding of basic rules and notation of logic

1

greatly facilitates communication of mathematical concepts as well as one's own mental organization of these concepts.

The study of formal logic dates back to the ancient Greeks. Such men as Thales and Aristotle gave dignity to man's power of reason as they formalized deductive logic. Later, such men as Euler, De-Morgan, and Boole gave definite structure to the study of logic and laid a logical foundation for other areas of mathematics.

The study of logic can be fascinating to the "logical mind" and frustrating to the "illogical mind." In fact, some become so frustrated in trying to train illogical minds to think logically that they share the feelings of Cal Craig, who asked, "How are you going to teach logic in a world where everybody talks about the sun setting, when it's really the horizon rising?" Still, believing that man is by nature a rational being, we begin our study of mathematics with a study of logic.

1.2 STATEMENTS AND NEGATIONS

In mathematics and the study of logic, we consider a *statement,* or *proposition,* to be a sentence that is either true or false, but not both. Another way of saying this is that a statement is a sentence with *truth value.*

Example

(i) "All roses are red" is a false statement.

(ii) "Mathematics is difficult for some people" is a true statement.

(iii) "This sentence is false" is not a statement since it can be both true and false.

(iv) "$5 + 4 = 9$" is a true statement.

(v) "$5 + 3 = 9$" is a false statement.

(vi) "$5 + n = 9$" is not a statement since it is neither true nor false; that is, its truth or falsity depends upon the replacement for the variable n.

Just as we use symbols, or variables, to represent numbers in general in algebra, we use symbols in much the same way in logic to represent statements when we wish to discuss statements in general rather than specific statements. For example, where we use p to represent a statement, we wish to define the *negation of p*, which we denote $\sim p$.

definition 1.1 (negation)

> For any statement p, a *negation* of p, denoted $\sim p$, is a statement that must be false when p is true and must be true when p is false (that is, a statement and its negation cannot both be true or both be false).

A common misconception about the concept of a negation is that one sentence is the statement and the other is the negation, when actually they are both statements, which are negations of each other. It is much like the fact that there are two sexes, which are opposite each other, rather than one being *the sex* and the other being the *opposite sex*.

Example

(i)   The statements "I like mathematics" and "I do not like mathematics" are negations of each other.

(ii) The statements "I am not an American" and "I am an American" are negations of each other.

(iii) The statements "I am a college graduate" and "I am not a college graduate" are negations of each other.

Statements such as those in the previous example are quite direct and easy to negate. There are statements, however, the negations of which are much more difficult to determine, particularly those involving one of the words "all," "none," or "some." For example, one may assume that the negation of the statement "All of you will pass this course" is "None of you will pass this course." This is not the case, however, since both statements can be false at the same time; that is some may pass the course, while some do not. Actually, the negation of any statement of the type "All do . . . " can be stated in many forms, the most common of which are "Some do not . . . " and "Not all do. . . ."

Example

> The negation of the statement "All of you will pass this course" can be stated in any of the following forms:

"Some of you will not pass this course."
"Not all of you will pass this course."
"It is not true that all of you will pass this course."

Just as "All do . . . " and "Some do not . . . " are negations of each other, "Some do . . . " and "None do . . . " are negations of each other.

Example

The statements "Some of you will pass this course" and "None of you will pass this course" are negations of each other. The statements "Some of you will pass this course" and "All of you will pass this course" are not negations since they can both be true at the same time.

For any statement p, we denote the negation of p by $\sim p$, and since they are negations of each other, we can say that the negation of $\sim p$ is p itself; that is,

$$\sim (\sim p) \leftrightarrow p.$$

Exercise Set 1.1
1. Identify each of the following as a statement or not a statement.
    (a) This book is blue.   (d) $X + 7 = 9$
    (b) Today is Tuesday.   (e) $5 + 4 = 3$
    (c) Good grief!
2. State the negation of each of the following statements.
    (a) I am broke.
    (b) I do not like school.
    (c) Dating is fun.
    (d) Marriage is not for everyone.
    (e) Married men live longer than single men.
    (f) All roses are red.
    (g) Some politicians are honest.
    (h) No one is happy in bondage.
    (i) Everyone likes mathematics.
    (j) Some animals cannot be tamed.
    (k) The Cardinals won the game.
    (l) $2x + 3$ is greater than 5.
3. For any statement p, when is $\sim p$ true?

1.3 CONJUNCTIONS AND DISJUNCTIONS

Just as we use both simple and compound sentences in our speech and writing, we use simple and compound statements in logic. Two connectives used to make compound statements from two or more simple statements are the words "and" and "or." The compound sentence formed when two simple statements are connected by the word "and" is called a *conjunction,* and the compound sentence formed when two simple statements are connected by the word "or" is called a *disjunction.* The connective "and" is often denoted by the symbol $\wedge$, while the connective "or" is denoted by the symbol $\vee$.

Example

If p represents the statement

"I will get married"

and q represents the statement

"I will graduate,"

then $p \wedge q$ represents the statement

"I will get married and I will graduate";

$p \vee q$ represents the statement

"I will get married or I will graduate";

$\sim p \wedge q$ represents the statement

"I will not get married and I will graduate";

(*Note:* $\sim p \wedge q$ represents $(\sim p) \wedge q$ rather than $\sim(p \wedge q)$)
$p \vee \sim q$ represents the statement

"I will get married or I will not graduate"; and

$\sim p \wedge \sim q$ represents the statement

"I will not get married and I will not graduate."

In general usage, the word "or" has two meanings – the *exclusive* and the *inclusive*. The exclusive "or" is used to mean one or the other, *but not both*. For example, "Whom do you wish to marry, Debbie *or* Sue?" The inclusive "or," however, means one or the other *or both*. For example, "Can you sing or dance?" In mathematics, we use the inclusive "or" unless otherwise specified.

Often, the word "but" is used in place of the connective "and" for the sake of better grammar and clarification, but this does not change the meaning of the conjunction, so the same symbol $\land$ is used.

Example

If p represents the statement

"I like mathematics"

and q represents the statement

"I like physics,"

then $p \land \sim q$ represents the conjunction

"I like mathematics *and* I do not like physics"

which may also be stated

"I like mathematics *but* I do not like physics."

In the previous section, we considered the process of negating simple statements. Now let us consider the process of negating compound statements. First, consider the negation for the conjunction $p \land q$, denoted $\sim(p \land q)$. Is this the same as $\sim p \land \sim q$? Before we answer the question in general, let us consider a specific example. If a job opening is listed for someone who is both a mathematician and an engineer and you hear a man say that he is not qualified for the job, that is, he is not both a mathematician and an engineer, does that necessarily mean that he is not a mathematician *and* he is not an engineer? No. It could mean that he is a mathematician but not an engineer, or an engineer but not a mathematician, or perhaps neither. In other words, either he is not a mathematician *or* he is not an engineer; that is,

$$\sim(p \wedge q) = \sim p \vee \sim q.$$

Now consider a similar ad asking for a person who is either a mathematician *or* an engineer. If you are not qualified for this position, then you are not a mathematician or an engineer. This does not mean that you are not a mathematician *or* you are not an engineer, but that you are not a mathematician *and* you are not an engineer; that is,

$$\sim(p \vee q) = \sim p \wedge \sim q.$$

Thus, to negate a conjunction or disjunction of two or more simple statements, we negate each of the simple statements and change all conjunctions to disjunctions and all disjunctions to conjunctions.

Example
(i) $\sim(p \wedge q) = \sim p \vee \sim q$
(ii) $\sim(\sim p \vee q) = p \wedge \sim q$
(iii) $\sim(p \vee \sim q) = \sim p \vee q$
(iv) $\sim(p \vee q \vee \sim r) = \sim p \wedge \sim q \wedge r$
(v) $\sim(\sim p \wedge \sim q \wedge r) = p \vee q \vee \sim r$

Exercise Set 1.2
1. If p represents the statement "I will try" and q represents the statement "I will succeed," write the following out in words.

(a) $p \wedge q$ (h) $\sim p \vee \sim q$
(b) $p \vee q$ (i) $\sim p \wedge \sim q$
(c) $p \wedge \sim q$ (j) $\sim(p \vee q)$
(d) $\sim p \wedge q$ (k) $\sim(p \wedge q)$
(e) $p \vee \sim q$ (l) $\sim(p \wedge \sim q)$
(f) $\sim p \vee q$ (m) $\sim(\sim p \vee q)$
(g) $\sim p \vee p$ (n) $\sim(\sim p \wedge \sim q)$

2. Let p represent the statement "I passed algebra" and q represent the statement "I passed English." Write each of the following symbolically.

(a) I passed algebra and failed English.
(b) I failed both algebra and English.
(c) I passed both algebra and English.
(d) I failed either algebra or English.
(e) I didn't fail either algebra or English.

(f) I didn't pass either algebra or English.

(g) Either I passed algebra and failed English or I passed English and failed algebra.

(h) I didn't pass both algebra and English.

(i) I didn't fail both algebra and English.

(j) Either I passed English or I failed English.

3. Identify each of the following compound statements as true or false, where p and q represent any simple statements. (*Note:* If the statement is not *always* true, then it should be labeled false.)

(a) $\sim(p \vee q) = \sim p \vee \sim q$

(b) $\sim(p \vee q) = \sim p \vee q$

(c) $\sim(p \wedge \sim q) = \sim p \vee q$

(d) $\sim(\sim p \wedge q) = p \vee \sim q$

(e) $\sim(p \vee \sim p) = \sim p \wedge p$

(f) $\sim(\sim p \vee \sim q) = p \wedge q$

1.4 CONDITIONAL AND BICONDITIONAL STATEMENTS

In the previous section, we discussed compound statements that are conjunctions or disjunctions, but not every compound statement fits into either of these categories. For example, we can take the two simple statements "I failed the test" and "I will fail the course" and form the conjunction, "I failed this test and I will fail the course," and the disjunction, "Either I failed this test or I will fail the course." We can also use the words "if" and "then" to form the compound statement "If I failed this test, then I will fail the course." Compound statements of the form "if p then q," where p and q are simple statements and "if" and "then" are either written or understood, are called *conditional statements,* and are denoted $p \rightarrow q$. In such statements, p is called the *hypothesis* and q is called the *conclusion.*

Example

(i) "If it rains, then we shall cancel the game."
Hypothesis: It rains.
Conclusion: We shall cancel the game.

(ii) "If you're late, we won't go."
Hypothesis: You're late.
Conclusion: We won't go.

(iii) "I'm leaving if you don't learn some manners."
Hypothesis: You don't learn some manners.
Conclusion: I'm leaving.

Now consider the conditional statement "If John is pitching, then his team is winning." Which of the following sentences have the same meaning?

1) If John's team is winning, then John is pitching.
2) If John is not pitching, then his team is not winning.
3) If John's team is not winning, then he is not pitching.
4) John is not pitching, or his team is winning.

Even though all of these sentences are similar to the original sentence, only the last two convey the same meaning or have the same truth value. All of them, however, are related, both to one another and to the original sentence "If John is pitching, then his team is winning." Sentence 1) is called the *converse* of this sentence; sentence 2) is called its *inverse;* and sentences 3) and 4) are just different forms of its *contrapositive.*

Just as every statement has a negation, every conditional statement has a *converse,* an *inverse,* and a *contrapositive,* defined as follows.

definition 1.2 (converse, inverse, and contrapositive)

Where $p \rightarrow q$ is any conditional statement,

(i) $q \rightarrow p$ is its *converse;*
(ii) $\sim p \rightarrow \sim q$ is its *inverse;* and
(iii) $\sim q \rightarrow \sim p$ is its *contrapositive.*

From the definition of converse, inverse, and contrapositive, we can see that the converse of a conditional statement is the conditional statement formed by *interchanging* the hypothesis and conclusion of the original statement; the inverse is formed by *negating* both the hypothesis and conclusion of the original statement; and the contrapositive is formed by *both* interchanging and negating the hypothesis and conclusion of the original statement. Thus, the contrapositive is often defined as the *converse of the inverse* or *the inverse of the converse.* Also note that the inverse and the converse are contrapositives of each other. These relationships are illustrated in Figure 1.1.

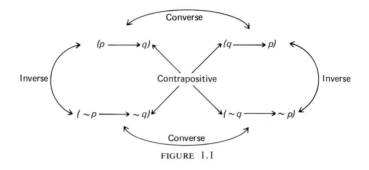

FIGURE 1.1

Example

Consider the conditional statement

"If you are a college student, then you are intelligent."
Hypothesis = You are a college student.
Conclusion = You are intelligent.

Its converse is:

"If you are intelligent, then you are a college student";

Its inverse is:

"If you are not a college student,
then you are not intelligent";

and its contrapositive is:

"If you are not intelligent,
then you are not a college student."

From the previous example, it is not difficult to see that the truth or falsity of a conditional statement in no way guarantee the truth or falsity of its converse or inverse. It may be less clear, however, that the truth or falsity of a conditional statement does guarantee the truth or falsity of its contrapositive. We say that a conditional statement and its contrapositive are *logically equivalent;* that is, they have the same truth value. From Figure 1.1 we see that the inverse and converse of a conditional statement are contrapositives of each other; therefore they, too, are logically equivalent.

Example

If the statement "If you are a Frenchman, then you are a European" is true, then so is its contrapositive "If you are not a European, then you are not a Frenchman." Also, if one is false, then so is the other. Likewise, the converse "If you are a European, then you are a Frenchman" and its contrapositive (which is the inverse of the original statement) "If you are not a Frenchman, then you are not a European" have the same truth value.

Many theorems and definitions in mathematics are stated in the form "p if and only if q," which is the same as saying "if q then p and if $\sim q$ then $\sim p$" or symbolically $(q \rightarrow p) \wedge (\sim q \rightarrow \sim p)$. Since $\sim q \rightarrow \sim p$ has the same truth value as $p \rightarrow q$, "p if and only if q" means $(q \rightarrow p) \wedge (p \rightarrow q)$, which is called a biconditional statement and is denoted

$$p \leftrightarrow q.$$

Not all conditional statements include the actual words "if" and "then" even though the meaning is there. For example, statements beginning with the word "all" or "none" can often be changed to "if p then q" statements without altering their meanings.

Example

(i)   The statement "All girls are pretty" can also be stated "If you are a girl, then you are pretty."

(ii) The statement "No hermits are happy" can also be stated "If you are a hermit, then you are not happy."

(iii) The statement "None of the seniors were invited to our party" can also be stated "If you are a senior, then you were not invited to our party."

The previous example illustrates how the wording of a statement may be changed without changing its meaning. We consider the two statements in each part of the example to be the same since they are merely different ways of expressing the same idea or concept. The ideas and concepts are the object of our concern in a study of logic, so the wording will often be changed to make the idea more understandable. For example, in determining the converse or contrapositive of a conditional statement, it often becomes necessary to change the tense of the verb or make other changes in the wording in order to retain the correct concept and at the same time to be grammatically correct.

Example

Conditional statement: If I pass this course, then I shall graduate.

Contrapositive: If I do not graduate, then I shall not have passed this course.

Converse: If I graduate, then I shall have passed this course.

Inverse: If I do not pass this course, then I shall not graduate.

Exercise Set 1.3

1. Express each of the following symbolically.
    (a) If you're a normal young man, then you like girls.
    (b) If you are married, then you have a mother-in-law.
    (c) If you are a freshman or a sophomore, then you are invited to the party.
    (d) All married students and only married students live off campus.
    (e) If you pass this class and pay your fees, then you will graduate.
    (f) $1x = 0$ if and only if $x = 0$.
    (g) If $xy = 0$, then $x = 0$ or $y = 0$.
    (h) $x + y = z + y$ if and only if $x = z$.
    (i) $xy < xz$ if and only if $y < z$ or $x < 0$.
    (j) If $xs < xz$ and $s > z$, then $x < 0$.

2. Write out the meaning of each of the following, where p represents the statement "I like mathematics" and q represents the statement "I like physics."
    (a) $\sim p \wedge \sim q \rightarrow \sim p$
    (b) $\sim p \vee q \leftrightarrow \sim (p \wedge \sim q)$
    (c) $\sim (p \vee q) \rightarrow \sim q$
    (d) $p \leftrightarrow \sim (\sim p)$
    (e) $p \wedge q \rightarrow p \vee q$

(f) $\sim(p \wedge q) \leftrightarrow \sim p \vee \sim q$

(g) $p \wedge q \rightarrow p$

(h) $(p \vee q) \wedge \sim p \rightarrow q$

(i) $\sim p \rightarrow \sim(p \wedge q)$

(j) $\sim p \wedge \sim q \leftrightarrow \sim(p \vee q)$

3. Identify each of the following as a converse, inverse, negation, contrapositive, or restatement of "All intelligent students like mathematics." If it is none of these, write "none."

(a) Not all intelligent students like mathematics.

(b) No intelligent students like mathematics.

(c) Some intelligent students like mathematics.

(d) Some intelligent students do not like mathematics.

(e) All intelligent students do not like mathematics.

(f) If you are an intelligent student, then you like mathematics.

(g) If you are not an intelligent student, then you do not like mathematics.

(h) If you like mathematics, then you are an intelligent student.

(i) If you do not like mathematics, then you are not an intelligent student.

(j) All people who like mathematics are intelligent students.

4. Find the converse, inverse, and contrapositive of each of the following.

(a) If you do well in class, then you will enjoy it.

(b) If school were easier, then I would graduate.

(c) If it rains, then we will cancel the football game.

(d) If I fail, then so will you.

(e) If I play football, then I can't play basketball.

(f) If I believed she would accept, then I would ask her to marry me.

(g) $\sim p \rightarrow \sim q$

(h) $\sim q \rightarrow p$

(i) $\sim q \rightarrow \sim p$

(j) $p \rightarrow \sim q$

1.5 TRUTH TABLES

We have discussed simple statements, negations, conjunctions, disjunctions, conditional statements, and biconditional statements

—all of which are statements, and a statement is defined as being either true or false but not both. Often tables, called *truth tables,* are used to explore the truth or falsity of these various types of statements. In fact, truth tables can actually be used to define these various compound statements. Such tables illustrate all possible combinations of truth values of simple statements involved in the truth values of the resulting compound statements. For example, consider Table 1.1, which is called the truth table for $p \wedge q$. There are four cases to consider. The first is the case in which p is true and q is true, with the result that $p \wedge q$ is true. The second case is that in which p is true and q is false, so $p \wedge q$ is false. In the third case, p is false and q is true, so $p \wedge q$ is again false. The final case is that in which p is false and q is false, so $p \wedge q$ is also false.

TABLE 1.1

	p	q	$p \wedge q$
Case 1	T	T	T
Case 2	T	F	F
Case 3	F	T	F
Case 4	F	F	F

Now contrast Table 1.1 with Table 1.2, which is the truth table for $p \vee q$. This comparison illustrates the difference in the meaning of the words "and" and "or." The only way in which $p \wedge q$ can be true is if p is true and q is also true. In any other case, $p \wedge q$ is false. In contrast, the only time $p \vee q$ is false is when p is false and q is false. Any other time, $p \vee q$ is true.

TABLE 1.2

p	q	$p \vee q$
T	T	T
T	F	T
F	T	T
F	F	F

Each of the connectives that we have discussed, as well as any combination of them, can be defined by a truth table. In each case, the truth table can also be constructed from the definitions of the connectives. Table 1.3 is a truth table combining the separate truth tables for $p \wedge q$, $p \vee q$, $p \rightarrow q$, $q \rightarrow p$, $p \leftrightarrow q$, $\sim p$, and $\sim q$.

TABLE 1.3

p	q	$p \wedge q$	$p \vee q$	$p \rightarrow q$	$q \rightarrow p$	$p \leftrightarrow q$	$\sim p$	$\sim q$
T	T	T	T	T	T	T	F	F
T	F	F	T	F	T	F	F	T
F	T	F	T	T	F	F	T	F
F	F	F	F	T	T	T	T	T

That which is most confusing about Table 1.3 to most students has to do with Cases 3 and 4 of $p \rightarrow q$ and Cases 2 and 4 of $q \rightarrow p$. The reason all of these are labeled "true" is that a conditional statement merely says, "*If* the hypothesis is true, then the conclusion is true." It says nothing of the case in which the hypothesis is false, so the conclusion cannot contradict such a case, regardless of what it is. Thus, we arbitrarily assign the truth value of "true" to every case in which the hypothesis is false. For example, if Susan makes the statement, "If Al takes Janet to the prom, then I'll never go with him again," then we accept her statement as true, whether or not she goes with Al again, unless Al does take Janet to the prom and Susan still goes with him again. After all, she didn't say that she would go with him again if he did not take Janet to the prom.

We say that $p \rightarrow q$ is true if q is true whenever p is true. The biconditional case is much more limited, however. We say that $p \leftrightarrow q$ only when p and q have exactly the same truth value; that is, if p is true, then q is true, and if p is false, then q is false.

Truth tables are often very helpful as orderly methods of considering possible outcomes of conditional statements and illustrating the relationship of one statement to another. For example, let us look at some of the problems of the previous exercise set in terms of truth tables.

Example

(i) Consider the statement $(p \wedge q) \rightarrow (p \vee q)$ and construct a truth table for the statement. (See Table 1.4.)

TABLE 1.4

p	q	$p \wedge q$	$p \vee q$	$p \wedge q \rightarrow p \vee q$
T	T	T	T	T
T	F	F	T	T
F	T	F	T	T
F	F	F	F	T

(ii) Construct a truth table of $\sim(p \wedge q) \leftrightarrow (\sim p \vee \sim q)$. (See Table 1.5.)

TABLE 1.5

p	q	$\sim p$	$\sim q$	$p \wedge q$	$\sim(p \wedge q)$	$\sim p \vee \sim q$	$\sim(p \wedge q) \leftrightarrow \sim p \vee \sim q$
T	T	F	F	T	F	F	T
T	F	F	T	F	T	T	T
F	T	T	F	F	T	T	T
F	F	T	T	F	T	T	T

(iii) Construct a truth table of $\sim(p \vee q) \leftrightarrow (\sim p \wedge \sim q)$. (See Table 1.6.)

TABLE 1.6

p	q	$\sim p$	$\sim q$	$p \vee q$	$\sim(p \vee q)$	$\sim p \wedge \sim q$	$\sim(p \vee q) \leftrightarrow \sim p \wedge \sim q$
T	T	F	F	T	F	F	T
T	F	F	T	T	F	F	T
F	T	T	F	T	F	F	T
F	F	T	T	F	T	T	T

Note that in all three tables of the previous example, the truth table for the conditional or biconditional statement consists of T's only. In other words, the statement is always true. When this is true, the statement is called a *tautology*. Whenever a conditional statement, $p \to q$, is a tautology, we call the relationship of p to q an *implication;* that is, p *implies* q, and denote it $p \Rightarrow q$. Remember that there is a difference in a conditional statement and an implication. The entire statement "If p then q" is the conditional statement, while the relationship of p to q, provided that $p \to q$ is always true, is the implication.

definition 1.3 (implication)

For any statements p and q, we say that p *implies* q, denoted $p \Rightarrow q$, if and only if $p \to q$ is a tautology, that is, is true in every case.

When a biconditional statement $p \leftrightarrow q$ is a tautology, then $p \Rightarrow q$ and $q \Rightarrow p$. Thus, we call the relationship between p and q an *equivalence* relation; that is, p and q are *equivalent* statements or are *logically equivalent*.

definition 1.4 (equivalent statements)

For any statements p and q, we say that p is *equivalent* to q, denoted $p \Leftrightarrow q$ or $p \equiv q$, if and only if the biconditional statement $p \leftrightarrow q$ is a tautology.

Example

Does $\sim p$ imply $\sim(p \lor q)$? To answer the question, we construct a truth table for $\sim p \rightarrow \sim(p \lor q)$. (See Table 1.7.)

TABLE 1.7

p	q	$\sim p$	$p \lor q$	$\sim(p \lor q)$	$\sim p \rightarrow \sim(p \lor q)$
T	T	F	T	F	T
T	F	F	T	F	T
F	T	T	T	F	F
F	F	T	F	T	T

Since the hypothesis is true and the conclusion is false in the third case, the conditional statement is false in this case so is not a tautology. Therefore, $\sim p$ does not imply $\sim(p \lor q)$.

Example

Does $\sim p$ imply $\sim(p \land q)$? Is $\sim p$ equivalent to $\sim(p \land q)$? (See Table 1.8.)

TABLE 1.8

p	q	$\sim p$	$p \land q$	$\sim(p \land q)$	$\sim p \rightarrow \sim(p \land q)$
T	T	F	T	F	T
T	F	F	F	T	T
F	T	T	F	T	T
F	F	T	F	T	T

This is a tautology, so $\sim p \Rightarrow \sim(p \land q)$; however, the tables for $\sim p$ and $\sim(p \land q)$ are not identical, so $\sim p$ is not equivalent to $\sim(p \land q)$.

Truth tables can also be constructed from three or more simple statements, such as p, q, and r. A truth table made up from 3 statements would have 8 cases to consider, and a table using 4 statements would have 16 cases to consider; thus they can become very cumbersome. Since this book is not designed for logic courses, we shall not even consider such tables. The purpose of the logic in this book is merely to aid in the understanding of the theorems and the mathematical development of the topics to be discussed in later chapters, and this need does not dictate a deeper or more thorough study of compound statements and truth tables.

Exercise Set 1.4

1. Construct a truth table for each of the following.
    (a) $\sim p \lor q$
    (b) $p \land \sim q$
    (c) $p \lor \sim p$
    (d) $\sim(p \land \sim q)$
    (e) $\sim(p \land q)$
    (f) $(p \lor q) \land \sim p$
    (g) $\sim(p \lor q)$
    (h) $(p \lor q) \lor \sim p$
    (i) $\sim(p \lor q) \lor \sim(p \land q) \to \sim(p \land q)$
    (j) $\sim p \lor \sim q \to \sim(p \lor q)$
    (k) $(p \to q) \leftrightarrow (\sim q \to \sim p)$
    (l) $(q \to p) \leftrightarrow (\sim p \to \sim q)$
    (m) $\sim(p \land q) \to \sim(p \lor q)$
    (n) $\sim(p \lor q) \to \sim(p \land q)$
    (o) $(p \lor q) \land \sim p \to q$
    (p) $[(p \to q) \land p] \to q$
    (q) $q \to (p \lor q)$
    (r) $(p \lor q) \to (p \land q)$
    (s) $(p \to q) \land (q \to p)$
    (t) $\sim(p \lor q) \to \sim p \lor \sim q$

2. Tell which of the statements in Problem 1 represent tautologies and which do not.

3. Identify each of the following as true or false. (Remember that "true" means *always* true.)
    (a) $(p \land q) \Rightarrow p$
    (b) $(p \lor q) \Rightarrow p$
    (c) $p \Rightarrow (p \land q)$
    (d) $p \Rightarrow (p \lor q)$
    (e) $\sim p \Rightarrow \sim(p \land q)$
    (f) $\sim p \Rightarrow \sim(p \lor q)$
    (g) $\sim(p \lor q) \Rightarrow \sim(p \land q)$
    (h) $\sim(p \land q) \Rightarrow \sim(p \lor q)$

(i) $(p \lor q) \land \sim q \Leftrightarrow p$

(j) $\sim (\sim p \land q) \Leftrightarrow (p \lor \sim q)$

(k) $(p \lor q) \Rightarrow (p \land q)$

(l) $(p \land q) \Rightarrow (p \lor q)$

(m) $\sim (p \land q) \Rightarrow (\sim p \land \sim q)$

(n) $\sim (p \lor q) \Leftrightarrow \sim (p \land q)$

(o) $\sim (p \lor q) \Leftrightarrow (\sim p \lor \sim q)$

(p) $\sim (p \lor q) \Rightarrow (\sim p \lor \sim q)$

(q) $(p \to q) \land p \Rightarrow q$

(r) $(p \to q) \Leftrightarrow (q \to \sim p)$

(s) $(p \land \sim q) \Rightarrow (\sim p \lor \sim q)$

(t) $\sim (p \to \sim q) \Leftrightarrow (\sim p \to q)$

1.6 DEDUCTIVE ARGUMENTS

A deductive argument is one in which one accepts basic asser-
tions, called *premises,* as being true and, on the basis of these premises, arrives at a *logical conclusion.* In a valid argument the truth of
the premises guarantees the truth of the conclusion. An argument can
easily be made using false premises. For example, one may accept as
premises the statements "Blondes are more intelligent than brunettes"
and "Marcia is a blonde" and validly conclude that "Marcia is more
intelligent than brunettes." Although the conclusion is *valid,* it is not
necessarily *true.* The conclusion may be false if either premise is
false, and that is highly possible since there is no assurance that
blondes are more intelligent than brunettes, nor can one be certain
that Marcia is really a blonde, in this day and age.

Actually, a valid deductive argument is merely an implication,
the conjunction of whose premises implies the conclusion. If the
conclusion is not implied by the premises, then we call the argument
invalid or a *fallacy.*

Example

Is the following a valid argument; that is, is the last statement a valid conclusion of those preceding it?

> If I am single, then I am free.
> I am single.
> Therefore, I am free.

solution

For some, the argument may seem obviously valid. For others, it may be necessary to construct a truth table to determine whether or not the premises imply the conclusion. To construct such a table, let the statement "I am single" be represented by p and "I am free," by q. In this way, we can symbolically represent the previous argument by

$$p \rightarrow q$$
$$\underline{p \qquad\qquad}$$
Therefore q (often denoted $\therefore q$)

or by

$$(p \rightarrow q) \wedge p \Rightarrow q.$$

The truth table for this argument was assigned as Problem 1(p) and 3(q) of the previous exercise set. Since the truth table represents a tautology, the implication holds, making the argument valid.

Example

The statements

"All seniors are snobs."
"Ron is a snob."
"Therefore, Ron is a senior."

do not constitute a valid argument since the first statement does not state that *only* seniors are snobs. To represent the argument symbolically, we first restate the first premise as the conditional statement, "If he is a senior, then he is a snob." We then represent the statement "He is a senior" by p, and "He is a snob" by q. Thus, the argument is

$$p \rightarrow q$$
$$\underline{q \qquad\qquad}$$
$$\therefore p$$

or $(p \rightarrow q) \wedge q \Rightarrow p$, which is a false statement, as illustrated in the following truth table (Table 1.9).

TABLE 1.9

p	q	$p \rightarrow q$	$(p \rightarrow q) \wedge q$	$(p \rightarrow q) \wedge q \rightarrow p$
T	T	T	T	T
T	F	F	F	T
F	T	T	T	F
F	F	T	T	T

Since the third case is false, the conditional statement $(p \rightarrow q) \wedge q \rightarrow p$ does not represent an implication, the statement $(p \rightarrow q) \wedge q \Rightarrow p$ is false, and therefore the argument is invalid.

Example

The following statements do not constitute a valid argument, even though the conclusion is true.

If n is a negative real number, then $|n| = -n$.
$|-5| = -(-5)$.
Therefore, -5 is a negative real number.

Let p represent the statement "n is a negative real number," and q represent the statement $|n| = -n$. Symbolically, the argument just presented becomes

$$p \rightarrow q$$
$$\underline{q\qquad}$$
$$\therefore p$$

which has the same truth table as the previous example.

Example

The following argument is valid even though the conclusion is not true.

If a, b, and c are whole numbers and $ac = bc$, then $a = b$.
5, 17, 0 are whole numbers and $5(0) = 17(0)$.
Therefore, $5 = 17$.

Symbolically:

$$p \wedge q \to r$$
$$\underline{p \wedge q}$$
$$\therefore r$$

As the preceding examples illustrate, arguments consisting of statements of the form

$$p \to q$$
$$\underline{p}$$
$$\therefore q \text{ (that is, } (p \to q) \wedge p \Rightarrow q)$$

or more generally

$$(p_1 \wedge p_2 \wedge p_3 \wedge \cdots \wedge p_n) \to q$$
$$\underline{(p_1 \wedge p_2 \wedge p_3 \wedge \cdots \wedge p_n)}$$
$$\therefore q$$

are valid arguments, while those of the form

$$p \to q$$
$$\underline{q}$$
$$\therefore p \text{ (that is, } (p \to q) \wedge q \Rightarrow p)$$

are invalid.

Another form of valid argument, commonly called the *law of syllogism*, is

$$p \to q$$
$$\underline{q \to r}$$
$$\therefore p \to r \text{ that is } (p \to q) \wedge (q \to r) \Rightarrow (p \to r)$$

Using this form of an argument, one can also obtain arguments using more than two premises. For example, we can obtain

$$p \to q$$
$$q \to r$$
$$\underline{p}$$
$$\therefore r$$

Example

Consider this argument:

Carol knows more about business than Brent does.
Brent knows more about business than Scott does.
Therefore, Carol knows more about business than Scott does.

The argument is valid and can be verified by the truth table for the law of syllogism.

Example

Consider the argument:

If I pass this exam, then I shall pass this course.
If I pass this course, then I shall have all of my group requirements filled.
If I have all of my group requirements filled, then I shall graduate.
I shall pass this exam.
Therefore, I shall graduate.

The argument is valid and can be illustrated symbolically as:

$$p \rightarrow q$$
$$q \rightarrow r$$
$$r \rightarrow s$$
$$\underline{p \qquad\quad}$$
$$\therefore s$$

We have shown, through several examples, that validity has little to do with truth. Their only relationship is that a valid argument guarantees the truth of a conclusion *whenever the premises are true*, but says nothing about the truth of a conclusion when the premises are false. This is the case whether we are arguing politics with a neighbor or presenting a proof of a mathematical theorem. The political arguments are seldom resolved due to the varied premises or assumptions upon which different people base their arguments. In mathematics, however, the premises we use are clearly established as either basic definitions, axioms (statements that we accept as true without proof), or previously proven theorems. We shall often refer to such premises as *properties*, whether previously proved or accepted without proof.

The reason the distinction between axioms and theorems will not always be made is that many properties that we accept as axioms at this level can be proved as theorems at a higher level of mathematics.

In proving theorems in mathematics, the most common methods used are merely logical arguments of the direct form

$$\begin{aligned} p &\to q \\ q &\to r \\ \hline \therefore p &\to r \end{aligned}$$

or the indirect form, using the contrapositive,

$$\begin{aligned} \sim r &\to q \\ q &\to \sim p \\ \hline \therefore \sim r &\to \sim p \text{ so } p \to r \end{aligned}$$

where p is the hypothesis of the theorem to be proved; r is its conclusion; and q is the union of the properties and definitions implied by p.

The purpose of proofs in this book is merely to aid understanding and clarify concepts. Thus, only proofs that tend to accomplish this task are given, while most theorems are stated without proof.

Exercise Set 1.5
1. Identify the final statement of each of the following arguments as a valid or an invalid conclusion of the premises.
   (a) If you work, then you will succeed.
       Dee works.
       Therefore, Dee will succeed.
   (b) If you are over thirty, then you are old.
       Tom is old.
       Therefore, Tom is over thirty.
   (c) All men are free.
       Jean is not a man.
       Therefore, Jean is not free.
   (d) Smoking is a harmful habit.
       Jack has a harmful habit.
       Therefore, Jack smokes.
   (e) No freshmen will be on this year's basketball team.
       Tim is no freshman.

Therefore, Tim will be on this year's basketball team.
(f) No mortal man is free of mistakes.
Some mortal men are good men.
Therefore, some good men are not free of mistakes.
(g) If McKay is intelligent and works hard, he will be promoted.
McKay is intelligent.
McKay does work hard.
Therefore, McKay will be promoted.
(h) Only intelligent girls become good wives.
Ann is not a good wife.
Therefore, Ann is not intelligent.
(i) Intelligent people do not get drunk.
Only drunk people sleep on the street.
Joe slept on the street last night.
Therefore, Joe is not intelligent.
(j) No one is perfect unless he is a competent mathematician.
All competent mathematicians like sports.
Jerry dislikes all sports.
Therefore, Jerry is imperfect.

2. Identify each of the following as a valid or an invalid argument.

(a) $p \rightarrow q$
$\underline{p \rightarrow r}$
$\therefore q \rightarrow r$

(b) $p \rightarrow r$
$\underline{q \rightarrow r}$
$\therefore p \rightarrow q$

(c) $p \rightarrow q$
$\underline{q \rightarrow \sim p}$
$\therefore p \rightarrow \sim p$

(d) $p \rightarrow q$
$\underline{\sim r \rightarrow \sim q}$
$\therefore p \rightarrow r$

(e) $\sim(p \wedge q)$
$\underline{p}$
$\therefore \sim q$

(f) $p \rightarrow q$
$\underline{p \vee q}$
$\therefore q$

(g) $p \vee q$
$p \vee \sim r$
$\underline{p \rightarrow r}$
$\therefore q \vee r$

(h) $p \rightarrow q$
$\underline{\sim q}$
$\therefore \sim q$

(i) $p \rightarrow q$
$\underline{q \wedge r}$
$\therefore \sim r \rightarrow \sim p$

3. Express each of the arguments in Problem 1 symbolically.
4. Find one valid conclusion to each of the following sets of prem-

ises, using all of the premises. If there are no valid conclusions, then so state.

(a) No politicians are honest.
     All farmers are honest.

(b) If a schoolteacher is honest, then he is not wealthy.
     Mr. Lowe is a wealthy schoolteacher.

(c) Mr. Vance will buy either stock or insurance.
     Mr. Vance will not buy insurance.

(d) I cannot sing and dance.
     I can dance.

(e) Anyone who majors in statistics should study some probability.
     Nancy is not majoring in statistics.

(f) If married men live longer than single men, then marriage is a healthy institution.
     Married men do not live longer than single men; it just seems that way.

(g) Cattlemen and forest rangers do not agree on anything.
     My father is a cattleman.
     My father and uncle agree on almost everything.

(h) All sorority girls are snobs.
     Bill goes with Janice.
     Bill will not go with a snob.

(i) All leaders have an obligation to set a good example.
     One who has the strength to stand by his convictions is a leader.
     Those who are obligated to set a good example grow through the experience.

(j) Jane has dated every man in her school who is tall, dark, *and* handsome.
     Mary has dated every man in the same school who is tall, dark, *or* handsome.
     Jane has never dated a man that Mary has dated.

(k) If I study enough, then I shall do well in all my classes.
     If I limit my social activity to weekends, then I shall study enough.
     If I stop bowling, then I shall limit my social activity to weekends.
     If I get cut off the bowling team, then I shall stop bowling.
     I do not do well in all of my classes.

1.7 SWITCHING NETWORKS

One way of forming a mental picture of compound statements is to consider such statements as *switching networks*. A switching network is an arrangement of wires and switches connecting two terminals in such a way that a *closed* switch permits the flow of current, while an *open* switch prohibits such a flow. Thus, we associate each simple statement p with a switch P such that a true statement is represented by a closed switch, and false statement is represented by an open switch. For example, Figure 1.2 illustrates the statement "p is true and q is false."

FIGURE 1.2

Figure 1.2 illustrates switches connected *in series*. When this is the situation, both P *and* Q must be closed for the current to flow from A to B; that is, p and q must be true. Thus, we see that *switches connected in series represent a conjunction.*

In addition to series, switches may be connected *in parallel* as in Figure 1.3. In this figure, the current can flow from A to B as long as either P or Q is closed; that is, if p or q is a true statement.

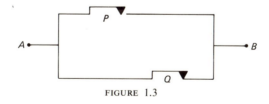

FIGURE 1.3

Thus, *switches connected in parallel can be used to represent a disjunction.*

Example

(i) The compound statement $p \wedge q \wedge r$ can be illustrated by a network of switches connected in the series in which the current can flow from A to B only if all three switches are closed. That is, the statement $p \wedge q \wedge r$ is true only if p is true, q is true, and r is true.

(ii) The compound statement $p \lor q \lor r$ can be illustrated by a network of switches connected in parallel. The current in this network can flow from A to B by one path or another provided that at least one of the switches is closed. Thus, the statement $p \lor q \lor r$ is true, provided that any one of the statements p, q, or r is true.

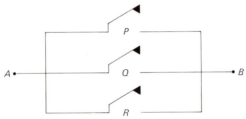

(iii) The compound statement $(p \land q) \lor r$ can be illustrated by the network in which the current can flow from A to B provided that either both P and Q are closed or that R is closed. Thus, the statement $(p \land q) \lor r$ is true whenever r is true or p and q are both true.

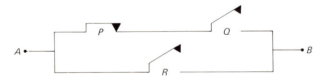

(iv) The compound statement $p \land (q \lor r)$ can be illustrated by the network in which the current can flow from A to B only if P is closed and either Q or R is also closed.

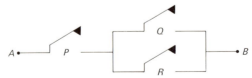

Just as different compound statements may be equivalent, so it is with switching networks. Thus, it is often possible to replace a switching network with one that is less complicated. For example, since the statement $(p \land q) \lor (p \land r)$ is equivalent to the statement $p \land (q \lor r)$, the network in Figure 1.4 is equivalent to the network in Part (iv) of the previous example.

In the same way, we find that the

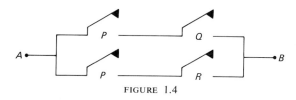

FIGURE 1.4

switching network in Figure 1.5 representing the statement

$$(p \land \sim q) \lor (p \land q) \lor (\sim q \land r)$$

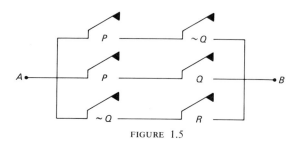

FIGURE 1.5

can be represented by the equivalent network in Figure 1.6, since the statement $(p \land \sim q) \lor (p \land q) \lor (\sim q \land r)$ is equivalent to the statement $p \lor (\sim q \land r)$. It is often easier to recognize equivalence of statements than equivalence of switching networks directly.

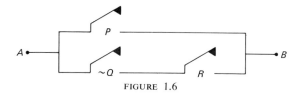

FIGURE 1.6

Example

The switching network

is equivalent to

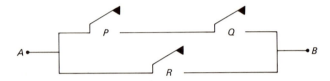

because the statement $(p \lor r) \land (q \lor r)$ is equivalent to the statement $(p \land q) \lor r$.

Exercise Set 1.6

1. Illustrate each of the following compound statements with a switching network.

    (a) $(p \land \sim q) \lor r$
    (b) $(\sim p \lor \sim q) \land r$
    (c) $(p \land \sim q) \lor (r \land q)$
    (d) $(p \lor \sim q) \land (r \lor q)$
    (e) $(\sim p \lor \sim q) \land (p \lor q)$
    (f) $r \land (p \lor \sim q)$
    (g) $(\sim p \lor q) \lor (p \lor \sim q) \lor \sim p$
    (h) $(p \land \sim q) \lor (\sim p \land q)$
    (i) $(p \lor q) \lor (r \land \sim p)$
    (j) $(p \land \sim q) \lor (\sim p \land q) \land (p \lor q)$

2. Write compound statements that correspond to the following networks.

    (a)

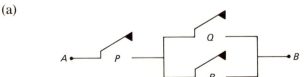

(b)

(c)

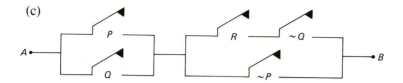

(d)

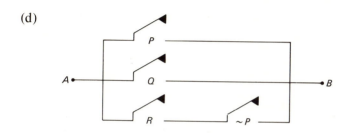

(e)

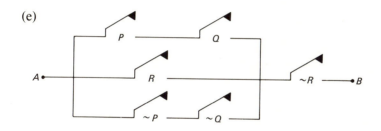

3. Use switching networks to determine and illustrate the truth value of each of the following statements.

(a) If p is true, then $(p \lor \sim q)$ is true.

(b) If p is true, then $(p \land \sim q)$ is true.

(c) If $(p \land \sim q)$ is true, then $(p \lor \sim q)$ is true.

(d) If $[(p \land q) \lor (p \land r)]$ is true, then p is true.

(e) If $(p \land q)$ is false, then $\sim p$ is true.

4. Design a network simpler than, but equivalent to, each of the following.

(a)

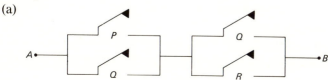

(b)

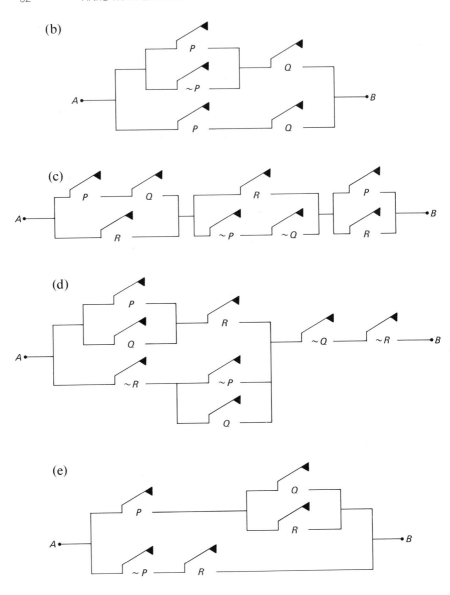

(c)

(d)

(e)

5. Design a network with three switches in which a light will turn on (the current will flow) if and only if *at least two* of the three switches are closed.

6. Design a network with three switches in which a light will turn on if and only if *exactly two* of the three switches are closed.

1.8 SUMMARY

In this chapter, we have defined basic terminology of logic and have discussed use of logical concepts in deductive arguments. We have also discussed truth tables as a method of analyzing the truth value of simple and compound statements as well as the validity of arguments. The following terms and symbols were introduced.

> Statement: A sentence that is either true or false but not both.
> Negation: Two statements that are negations cannot both be true or both be false; that is, if p is true, $\sim p$ is false, and if p is false, $\sim p$ is true.
> Disjunction $p \vee q$: p is true *or* q is true, or both.
> Conjunction $p \wedge q$: p and q.
> Conditional Statement $p \to q$: If p then q.
> Biconditional Statement $p \leftrightarrow q$: p if and only if q, that is, $p \to q$ and $q \to p$.
> Implication $p \Rightarrow q$: The relationship of p to q when $p \to q$ is a tautology.
> Equivalence $p \Leftrightarrow q$: The relationship between p and q when $p \leftrightarrow q$ is a tautology.
> Converse: The converse of $p \to q$ is $q \to p$.
> Inverse: The inverse of $p \to q$ is $\sim p \to \sim q$.
> Contrapositive: The contrapositive of $p \to q$ is $\sim q \to \sim p$.

Logical arguments can be investigated through the mediums of truth tables, but they can also be investigated symbolically according to their form. If the argument determines an implication, then it is valid. These arguments may be expressed in the form of an implication such as $(p \to q) \wedge p \Rightarrow q$ or in the form of a table

$$
\begin{array}{l}
p \to q \\
\underline{p } \\
\therefore q
\end{array}
$$

The following implications are the major valid argument forms; that is, any argument that is in one of these forms is valid.

$(p \vee q) \wedge \sim p \Rightarrow q$

$p \Rightarrow (p \vee q)$

$$(p \wedge q) \Rightarrow p$$
$$(p \rightarrow q) \wedge p \Rightarrow q$$
$$(p \rightarrow q) \wedge {\sim} q \Rightarrow {\sim} p$$
$$(p \rightarrow q) \wedge (q \rightarrow r) \Rightarrow (p \rightarrow r)$$
$$(p \wedge q) \Rightarrow (p \vee q)$$
$${\sim}({\sim} p) \Leftrightarrow p$$
$${\sim}(p \wedge q) \Leftrightarrow ({\sim} p \vee {\sim} q)$$
$${\sim}(p \vee q) \Leftrightarrow ({\sim} p \wedge {\sim} q)$$
$$(p \leftrightarrow q) \Leftrightarrow (p \rightarrow q) \wedge (q \rightarrow p)$$
$$(p \rightarrow q) \Leftrightarrow ({\sim} q \rightarrow {\sim} p)$$
$$(p \wedge q) \Leftrightarrow (q \wedge p)$$
$$(p \vee q) \Leftrightarrow (q \vee p)$$
$$(p \wedge q) \wedge r \Leftrightarrow p \wedge (q \wedge r)$$
$$(p \vee q) \vee r \Leftrightarrow p \vee (q \vee r)$$
$$p \wedge (q \vee r) \Leftrightarrow (p \wedge q) \vee (p \wedge r)$$
$$p \vee (q \wedge r) \Leftrightarrow (p \vee q) \wedge (p \vee r)$$

Review Exercise Set 1.7

1. State the negation of each of the following statements.
    (a) This world will never be free of war.
    (b) All women spend more money than their husbands earn.
    (c) No sport is more popular than football in America today.
    (d) Some governments are basically corrupt.
    (e) Some people do not care very much about anything.

2. State the converse, inverse, and contrapositive of each of the following conditional statements.
    (a) If you smoke, then you contribute to air pollution.
    (b) If you sit on the front row and smile, then you will get an A for the class.
    (c) If you enjoy your job, then you are lucky.
    (d) All intelligent people are basically honest.
    (e) If you are angry, then you are temporarily insane.

3. Construct a truth table for each of the following and tell which represent tautologies.
    (a) $p \wedge {\sim} q$ (d) $[(p \rightarrow q) \wedge {\sim} q] \rightarrow p$
    (b) ${\sim} p \vee q$ (e) $[(p \wedge q) \vee {\sim} p] \leftrightarrow q$
    (c) $(p \vee q) \rightarrow p$

4. Identify each of the following arguments as valid or invalid. If the

argument is valid, identify the form of the argument from those listed in the chapter summary.

(a) No one is successful without being dedicated.
     Henry is successful.
     Therefore, Henry is dedicated.

(b) If you enjoy your work, then you are good-natured.
     Tom is not good-natured.
     Therefore, Tom does not enjoy his work.

(c) All good men are honest.
     Jed is not a good man.
     Therefore, Jed is not honest.

(d) If you are patriotic, then you are concerned about ecology.
     Ken is concerned about ecology.
     Therefore, Ken is patriotic.

(e) $(a < b) \rightarrow (a + m = b)$
     $(a + m = b) \rightarrow (a + m + c = b + c)$
     $(a + m + c = b + c) \rightarrow (a + c + m = b + c)$
     $\underline{(a + c + m = b + c) \rightarrow (a + c < b + c)}$
     $\therefore (a < b) \rightarrow (a + c < b + c)$

chapter

2
SETS

2.1 INTRODUCTION

During the latter part of the nineteenth century, a German mathematician by the name of George Cantor began using the word "set" to describe a mathematical concept. His use of the word became common among other mathematicians, and finally an abstract branch of mathematics called *set theory* was developed. Although set theory was an advanced branch of mathematics, it was found that many of the ideas and terminology of set theory could be used to explain mathematical ideas at all levels of mathematics. Since that time, concepts associated with sets and set theory have appeared at all levels of mathematics from kindergarten to graduate school. In fact, the set concept has become the one most singularly identifying characteristic distinguishing "modern mathematics" from "traditional mathematics," even though set theory is not that new. Since the concept of a set precedes the concept of a number, it becomes difficult to study mathematics at any level without some understanding of set concepts and terminology. So, even though we cannot make a detailed study of set theory at this level of mathematics, we find it necessary to define and discuss some basic set concepts that are used in the discussion of other topics throughout the book.

2.2 SETS AND SET RELATIONS

The word "set" is an undefined term, but it is used to represent much the same idea in mathematics as in any other context. For example, a set of dishes, a set of furniture, a set of people, or a set of ideas fit very well into the mathematical concept of a set. In mathematics, we use the word *set* to represent any collection of "objects" (either concrete objects or abstract ideas) that is *well defined*. We use the phrase "well defined" to mean that the set is described or defined in such a way that we can determine whether or not any given object belongs to that set.

Example
The following are sets.
(i)   the set of all women;
(ii) the set of all past presidents of the United States;
(iii) the set of all whole numbers between 5 and 9.

Example
The following are not well defined, so are not sets:
(i)   three wealthy men;
(ii) all pretty women (there is no clear definition of *pretty women*);
(iii) all well-known authors.

Every set is a collection of *elements* or *members* of that set. When a set does not contain a great number of elements, a useful method of denoting the set is called the *tabulation* or *roster* method, in which the elements are all listed within set braces { } and separated by commas. For example, the set of all whole numbers between 5 and 9 can be denoted by

$${6,7,8}.$$

The tabulation method of denoting a set can also be used when a set has a great number of elements or even an infinite number of elements, provided that those elements are *ordered*. For example, the set of whole numbers less than 100 can be denoted

$${0,1,2, \ . \ . \ . \ ,99}$$

by listing at least the first three elements, followed by three dots and the final element. If we use the same notation except for the final element, then we denote a set with no final element. For example,

$${0,1,2, \ . \ . \ . \ }$$

denotes the set of *all* whole numbers, which is an infinite set.

When a set has neither few enough elements for tabulation to be convenient nor elements that are ordered, we normally use the *descriptive method* of naming the set. This method is also called *set builder notation*. Using this method, we use set braces to denote "the set containing," a variable x or n or whatever is preferred to denote "all x," a vertical bar to denote "such that," and then a description that every element of the set must satisfy.

$${x \mid x \text{ is a whole number between 5 and 9}} = {6,7,8}$$

the set containing   all elements x such that   each element x

Whether we use the tabulation method or set builder notation to denote a set, it would be very awkward to make several references to the same set. For example, we could say, "{1,2,3} is a set containing 3 elements. 1 is an element of {1,2,3}; 2 is an element of {1,2,3}; and 3 is an element of {1,2,3}." In order to avoid the necessity of repeatedly tabulating or describing a set, we use capital letters to denote sets and small letters to denote elements. We also use the symbol ∈

to denote the phrase "is an element of." Thus, we can say, "If $A = \{1,2,3\}$, then A is a set containing 3 elements. $1 \in A$, $2 \in A$, and $3 \in A$."

Often, sets are related by what is called a *subset* relation. This is to say that all of the elements of one set are also elements of the other set.

definition 2.1 (subset and superset)

For any sets A and B, A is a *subset* of B, denoted $A \subset B$, if and only if every element of A is an element of B; that is, $A \subset B$ if and only if for any element x,

$$x \in A \Rightarrow x \in B.$$

If $A \subset B$, then we say that B *contains* A or B is a *superset* of A, denoted $B \supset A$.

Example

(i) If $A = \{1,2,3\}$ and $B = \{0,1,2,3,4\}$, then $A \subset B$, which may also be denoted $B \supset A$.

(ii) If $C = \{a,b,c,d\}$ and $D = \{b,d\}$, then $D \subset C$, which may also be denoted $C \supset D$.

(iii) If $A = \{x \mid x$ is a senior at Jackson High School$\}$ and $B = \{y \mid y$ is a senior girl at Jackson High School$\}$, then $B \subset A$ and $A \supset B$.

A set may have many subsets. For example, if $A = \{0,1,2\}$, then certainly each of the sets

$$\{0\}, \{1\}, \{2\}, \{0,1\}, \{0,2\} \text{ and } \{1,2\}$$

is a subset of A, but what of $\{0,1,2\}$? Is A a subset of A? All that is necessary is to refer to the definition of a subset. If $x \in \{0,1,2\}$, then $x \in A$; therefore, $\{0,1,2\} \subset A$. In fact, $x \in A \Rightarrow x \in A$ for any element x and any set A; *therefore $A \subset A$ for every set A.*

Now let us consider a set with no elements, which we call the *null set* or *empty set* and denote it by $\varnothing$ or $\{ \}$. Certainly, this set exists, and most of us experience it as the set of all cash left at the end of the month, or the set of all women presidents of the United

States, or perhaps the set of all men who have proposed to a particular young lady. Is $\varnothing$ a subset of A where $A = \{0,1,2\}$? Since $x \in \varnothing$ is never true, $x \in \varnothing \Rightarrow x \in A$ is always true for any set A. Thus, there are eight subsets to $\{0,1,2\}$. They are $\varnothing$, $\{0\}$, $\{1\}$, $\{2\}$, $\{0,1\}$, $\{0,2\}$, $\{1,2\}$, and $\{0,1,2\}$. In general, there are 2^n subsets to a set with n elements, where n is any whole number.

Another important set relation is that of *equality*. We wish to define equality in such a way that neither the order in which elements appear nor the number of times any one element appears will affect the relation. For example, if a roll is passed around a class for students to sign, the result will be a tabulation of the set of students present in the class, regardless of the order in which the students sign the roll. If one student happens to sign the roll twice, this certainly does not make an extra person in the class—one of the elements is merely named twice. We use the subset relation to define equality in the following way.

definition 2.2 (equality of sets)
>For any sets A and B, we say that A *is equal to* B, denoted $A = B$, if and only if $A \subset B$ and $B \subset A$; that is, $A = B$ if

$$x \in A \Longleftrightarrow x \in B$$

>for any element x.

Example

$$\{1,2,3\} = \{1,3,2\} = \{3,1,2\}$$

Whether referring to the relation "is an element of," "is a subset of," "contains," or "is equal to," when we wish to state that the relation does *not* hold, we denote this by a slant bar through the relation symbol; that is,

"is not an element of" is denoted by $\notin$;
"is not a subset of" is denoted by $\not\subset$;
"does not contain" is denoted by $\not\supset$; and
"is not equal to" is denoted by $\neq$.

Whenever we describe a set, we are speaking within a certain frame of reference. For example, if one refers to the set of all numbers less than 3, to what set is he actually referring? Is it {1,2}, {0,1,2}, {2,1,0,−1,−2, . . . }, or perhaps the set of all rational numbers less than 3? This all depends on his frame of reference – whether he is thinking of natural numbers, whole numbers, integers, rational numbers, or perhaps real numbers. At any rate, the set to which he is referring is not well defined until the frame of reference, which we call the *universe,* denoted *U*, is described. Often, we illustrate sets with Venn diagrams, named after the English mathematician John Venn, in which we illustrate the universe by the interior of a rectangle and the set or sets in question by circles, as in Figure 2.1.

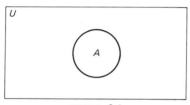

FIGURE 2.1

In Figure 2.1, note that the entire interior of the rectangle (including *A*) is represented by *U*. *A* represents only the interior of the circle. Up to this point, however, we have no way of naming the set that is represented in the diagram as that part of the interior of the rectangle that is not on the interior of the circle, that is, that which is within the universe but *not* in *A*. The set represented by this portion of the figure is called the *complement* of *A*.

definition 2.3 (complement of a set)
> For any set *A* in universe *U*, the *complement* of *A*, denoted $\bar{A}$, is the set of all elements of *U* that are *not* in *A*.

Example
> (i)   If $U = \{0,1,2, . . . ,10\}$ and $A = \{1,2,3\}$, then $A = \{0,4,5,6,7,8,9,10\}$.
> (ii) If *U* is the set of all students at U.C.L.A. and *A* is the set of all graduate students, then $\bar{A}$ is the set of all undergraduate students at U.C.L.A.

For the sake of convenience, most of the sets that we shall use in discussing the various set concepts in this chapter are sets of numbers. The following sets are those that will be used as universes most often.

$N = \{1,2,3, \ldots \}$ (the set of natural numbers)
$W = \{0,1,2, \ldots \}$ (the set of whole numbers)
$I = \{ \ldots ,-2,-1,0,1,2, \ldots \}$ (the set of integers)
$Q = \left\{ \dfrac{a}{b} \,\middle|\, a \in I \text{ and } b \in I \text{ and } b \neq 0 \right\}$ (the set of rational numbers)
$R = \{x \mid x \text{ is a rational number or } x \text{ is an irrational number}\}$ (the set of real numbers).

Another type of set in common usage in mathematics is called a solution set or replacement set. Such a set is used in connection with an open sentence, that is, a sentence containing a variable. The solution set of an open sentence is the set of all elements of the universe that can replace the variable and cause the open sentence to become a true statement.

Example

If W is the universe, the solution set to the sentence $x + 3 < 10$ ($<$ denotes "is less than") is

$$\{0,1,2,3,4,5,6\}.$$

Actually, the solving of equations and inequalities in algebra is determining the solution sets of open sentences.

Exercise Set 2.1
1.   Given that $A = \{0,1,2,3\}$, and $B = \{1,2,3\}$, and $U = \{0,1,2, \ldots ,10\}$, label each of the following statements as true or false.

(a) $3 \in A$

(b) $\{3\} \in A$

(c) $3 \subset A$

(d) $\{3\} \subset A$

(e) $A \subset B$

(f) $A \supset B$

(g) $B \subset B$

(h) $\varnothing \subset A$

(i) $\varnothing \in A$

(j) $\{\varnothing\} = \{0\}$

(k) $\{\varnothing\} = \varnothing$

(l) $\bar{A} \subset \bar{B}$

(m) $\bar{\varnothing} = U$

(n) $\bar{U} = \varnothing$

(o) $\bar{B} \subset U$

2. List all of the subsets of each of the following sets.
    (a) {0}   (d) ∅
    (b) {0,1}   (e) {∅}
    (c) {a,b,c}
3. Given $U = \{0,1,2, \ldots ,10\}$, tabulate each of the following sets.
    (a) $\{x \mid x$ is greater than 3 and less than 8$\}$
    (b) $\{x \mid x$ is greater than 3 or less than 8$\}$
    (c) $\{x \mid x$ is a multiple of 3$\}$
    (d) $\{x \mid 2x = 2 + x\}$
    (e) $\{x \mid x = 2x\}$
    (f) $\{x \mid x$ is greater than 5 and less than 4$\}$
    (g) $\{x \mid x$ is greater than 5 or less than 4$\}$
    (h) $\{x \mid x + x = 2x\}$
    (i) $\{x \mid x$ is less than 2x$\}$
    (j) $\{x \mid x$ is greater than 2x$\}$
4. Illustrate the relationship of A to B where $A \subset B$, using a Venn diagram.
5. Illustrate the relationship of A to B where $A \supset B$, using a Venn diagram.

2.3 SET OPERATIONS — UNION AND INTERSECTION

Since numbers are abstract while sets can be concrete, arithmetic operations on numbers are defined in terms of operations on sets. There are three set operations which we discuss in this book — two in this section and one in the following section.

In set theory, we use the word "intersection" in much the same way as it is used in other areas. The intersection of sets is that which the sets have in common. We define intersection in terms of the connective "and," using a symbol ∩ for intersection, which is similar to the logical symbol ∧ for "and."

definition 2.4 (intersection of sets)

For any sets A and B, the *intersection* of A and B, denoted $A \cap B$, is the set of all elements that are elements of both A and B. Symbolically, we say $A \cap B = \{x \mid (x \in A) \land (x \in B)\}$. If $A \cap B = \emptyset$, then we say that A and B are *disjoint* sets.

Example

(i)   If $A = \{0,1,2,3\}$ and $B = \{2,3,4\}$, then $A \cap B = \{2,3\}$.
(ii) If $C = \{A,B,\varnothing\}$ and $D = \{\varnothing,F,G\}$, then $C \cap D = \{\varnothing\}$.
(iii) If $F = \{a,b,c\}$ and $G = \{11,12,13\}$, then $F \cap G = \varnothing$.
(iv) If $A \subset B$, then $A \cap B = A$.

The intersection of sets can be illustrated with Venn diagrams. In Figures 2.2, 2.3, and 2.4, the shaded area represents the intersection of the given sets.

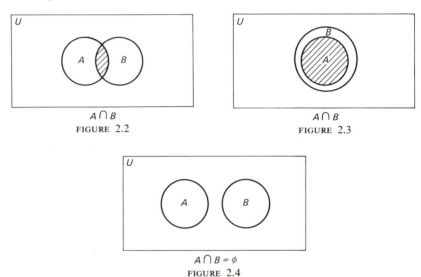

$A \cap B$
FIGURE 2.2

$A \cap B$
FIGURE 2.3

$A \cap B = \phi$
FIGURE 2.4

The second set operation that we define is that of *union*. Again, the word means much the same in mathematics as in common usage. Just as people who unite their resources have all of the resources of either or both at their disposal, the union of two sets is a set containing all of the elements contained in either or both sets. Thus, the operation "union" is defined in terms of the connective "or" and is represented by the symbol $\cup$, which is similar to the symbol $\vee$.

definition 2.5 (union of sets)

For any sets A and B, the union of A and B, denoted $A \cup B$, is the set of all elements, each of which is an element of A *or* an element of B. Symbolically,

$$A \cup B = \{x \mid (x \in A) \vee (x \in B)\}.$$

Example

(i)   If $A = \{0,1,2,3\}$ and $B = \{2,3,4\}$, then $A \cup B = \{0,1,2,3,4\}$.

(ii)   If $C = \{A,B,\varnothing\}$ and $D = \{\varnothing,F,G\}$,   then $C \cup D = \{A,B,F,G,\varnothing\}$.

(iii) If $F = \{a,b,c\}$ and $G = \{11,12,13\}$,   then $F \cup G = \{a,b,c,11,12,13\}$.

The operation of union can also be illustrated with Venn diagrams. The shaded areas of Figures 2.5, 2.6, and 2.7 represent the respective unions of sets.

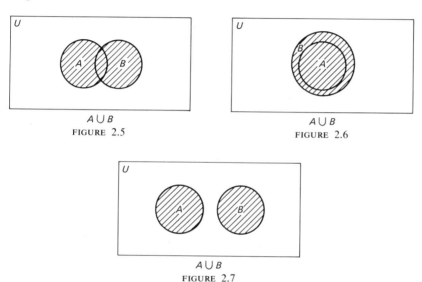

$A \cup B$

FIGURE 2.5

$A \cup B$

FIGURE 2.6

$A \cup B$

FIGURE 2.7

Often, the operations of union and intersection are used in conjunction with each other. Since each is a binary operation, only two sets can be operated on at a time, so parentheses are used to punctuate and designate which sets are operated on first. For example, to perform the operation $(A \cup B) \cap C$, we first find the set $A \cup B$, and then intersect this set with C. This is illustrated in Figure 2.8, in which lines are drawn one direction in $A \cup B$ and another direction in C. Since the last operation performed is intersection, and the intersection of two sets consists of those elements that the sets have in common, only the crosshatched area represents $(A \cup B) \cap C$. On the other hand, to perform the operations $A \cup (B \cap C)$, we first find the set $B \cap C$ and unite this set with A, as illustrated in Figure 2.9.

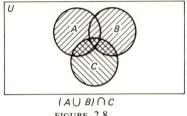

$(A \cup B) \cap C$

FIGURE 2.8

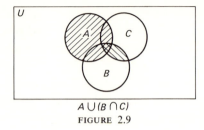

$A \cup (B \cap C)$

FIGURE 2.9

Since union is the last operation performed and the union of two sets is the set of all elements that are contained in one set or the other or both, the set $A \cup (B \cap C)$ is represented by the entire area that has lines running one direction or the other or both.

Example

(i) If $A = \{1,2,3,4,5\}$, $B = \{3,4,5,6\}$, and $C = \{2,4,6,8\}$, then
$(A \cup B) \cap C = \{1,2,3,4,5,6\} \cap \{2,4,6,8\} = \{2,4,6\}$
and
$A \cup (B \cap C) = \{1,2,3,4,5\} \cup \{4,6\} = \{1,2,3,4,5,6\}$

(ii) If $A = \{0,1,2,3,4\}$, $B = \{1,2,8,9\}$, $C = \{2,4,6,8\}$, and $U = \{0,1,2, \ldots ,10\}$, then
$$A \cup (\bar{B} \cap C) = \{0,1,2,3,4\}$$
$$\cup (\{0,3,4,5,6,7,10\} \cap \{2,4,6,8\})$$
$$= \{0,1,2,3,4\} \cup \{4,6\}$$
$$= \{0,1,2,3,4,6\};$$

and
$$(A \cup \bar{B}) \cap C = (\{0,1,2,3,4\}$$
$$\cup \{0,3,4,5,6,7,10\}) \cap \{2,4,6,8\}$$
$$= \{0,1,2,3,4,5,6,7,10\} \cap \{2,4,6,8\}$$
$$= \{2,4,6\}.$$

(iii) If $A = \{0,1,2\}$, $B = \{2,4,6\}$, $C = \{1,4,7\}$, and $U = \{0,1,2, \ldots ,10\}$, then
$$(\overline{A \cup B}) \cap C = \overline{\{0,1,2,4,6\}} \cap \{1,4,7\}$$
$$= \{3,5,7,8,9,10\} \cap \{1,4,7\}$$
$$= \{7\}.$$

The intersection of one set with the complement of another set as illustrated in Figure 2.10 is often called the *difference* of the sets, but this operation should not be confused with the operation of subtraction of numbers.

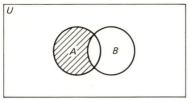

FIGURE 2.10

definition 2.6 (difference)

For any sets A and B, the difference of A and B, denoted $A - B$, is the set $A \cap \bar{B}$.

Example

If $A = \{2,4,6\}$, $B = \{4,5,6\}$, and $U = \{1,2,3, \ldots ,10\}$, then $A - B = A \cap \bar{B} = \{2,4,6\} \cap \{1,2,3,7,8,9,10\} = \{2\}$.

We have defined several set concepts, which correspond to concepts of logic. For example,

Complement ($\bar{A}$) corresponds to *negation* ($\sim p$).
Union ($A \cup B$) corresponds to *disjunction* ($p \vee q$).
Intersection ($A \cap B$) corresponds to *conjunction* ($p \wedge q$).

Because of these correspondences, an understanding of the logical terms can greatly facilitate understanding of set concepts, and each logical tautology proves a property concerning sets. Such properties will be investigated in the following exercise set and discussed in Section 2.5.

Exercise Set 2.2

1. Given that $A = \{2,3,5,7\}$, $B = \{1,3,5,7,9\}$, $C = \{2,4,6,8,10\}$, and $U = \{0,1,2, \ldots ,10\}$, find the following.

   (a) $A \cup B$
   (b) $A \cap B$
   (c) $A \cup \bar{B}$
   (d) $\bar{A} \cap B$
   (e) $A - B$
   (f) $(A \cup B) \cup C$
   (g) $(A \cup B) \cap C$
   (h) $B - \bar{A}$
   (i) $A \cap (B \cap C)$
   (j) $\bar{A} \cup (B \cap \bar{C})$
   (k) $\bar{A} \cap (B \cup \bar{C})$
   (l) $A \cup (B \cap \bar{C})$
   (m) $(\overline{A \cup B}) \cap C$
   (n) $A \cap (\overline{B \cap C})$
   (o) $(\bar{A} \cap \bar{B}) \cap C$
   (p) $(\bar{A} \cap \bar{B}) \cap (\overline{A \cup B})$

(q) $(\bar{A} \cup \bar{B}) \cap (\overline{A \cap B})$ (s) $(A \cup \bar{B}) \cup (B \cup \bar{A})$

(r) $A - (B \cap C)$ (t) $(\bar{C} \cap B) \cup (\bar{B} \cap C)$

2. Illustrate A, B, and C as three overlapping sets as in Figure 2.9 and illustrate the following operations by shading the appropriate areas.

(a) $(A \cup B) \cup C$ (i) $\overline{(A \cap B)} \cap C$

(b) $(A \cap B) \cap C$ (j) $(\bar{A} \cap B) \cap C$

(c) $A \cap (B \cup C)$ (k) $(A \cup B) \cap (A \cup C)$

(d) $(A \cap B) - C$ (l) $(A \cap B) \cup (A \cap C)$

(e) $\bar{A} \cap \bar{B}$ (m) $(A \cup \bar{B}) \cup \bar{C}$

(f) $A \cup (\bar{B} \cap C)$ (n) $A \cup (B \cup C)$

(g) $A \cup (\overline{B \cap C})$ (o) $(\bar{A} \cap B) \cap (\bar{A} \cup \bar{B})$

(h) $\bar{A} - (\bar{B} \cup C)$

3. For any sets A and B in universe U, identify each of the following statements as true or false.

(a) $(A \cup B) \subset (A \cap B)$ (f) $(\bar{A} \cap \bar{B}) = (\overline{A \cap B})$

(b) $(A \cap B) \subset (A \cup B)$ (g) $(\bar{A} \cup \bar{B}) = (\overline{A \cup B})$

(c) $A \subset (A \cup B)$ (h) $(\bar{A} \cap \bar{B}) = (\overline{A \cup B})$

(d) $(A \cap B) \subset A$ (i) $(\bar{A} \cup \bar{B}) = (\overline{A \cap B})$

(e) $A \subset (A \cap B)$ (j) $(\overline{A \cap B}) \subset (\overline{A \cup B})$

4. Where A, B, and C are sets illustrated in Figure 2.11, illustrate the following as shaded areas of Venn diagrams.

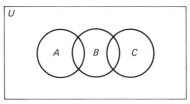

FIGURE 2.11

(a) $\bar{A} \cup \bar{C}$ (i) $\bar{B} \cup C$

(b) $\bar{A} \cap \bar{C}$ (j) $(\overline{A \cap B})$

(c) $(A \cap B) \cup C$ (k) $A - B$

(d) $(A \cup B) \cap C$ (l) $A - C$

(e) $(\overline{A \cap B}) \cup C$ (m) $(A \cup B) - C$

(f) $(A \cap B) \cup (B \cap C)$ (n) $(A - B) \cup C$

(g) $(A \cap C) \cup B$ (o) $(A \cap B) - \bar{C}$

(h) $A \cup \bar{C}$

5. Find the following for any sets A and B in universe U.

(a) $A \cap \bar{A}$

(b) $A \cup \bar{A}$

(c) $A \cap \varnothing$

(d) $A \cup \varnothing$

(e) $A \cap U$

(f) $(A \cup B) \cap (A \cap B)$

(g) $(A \cup B) \cup (A \cap B)$

(h) $\bar{B} \cap \varnothing$

(i) $A \cap B$, given that $A \subset B$

(j) $A \cup B$, given that $A \subset B$

2.4 SET OPERATIONS — CARTESIAN PRODUCT

When two sets are united or intersected, the resulting set has the same "kind" of elements as the set being operated on; that is, if A and B are sets of people, then $A \cup B$ and $A \cap B$ are sets of people, and if A and B are sets of numbers, then $A \cup B$ and $A \cap B$ are sets of numbers. Our third set operation, however, does not have this characteristic. If A and B are sets of numbers, then the *Cartesian product* of A and B, denoted $A \times B$, is a set of ordered pairs, that is, pairs (m,n) of elements, one, m, of which is designated as its *first component* and the other, n, as its *second component*.

definition 2.7 (Cartesian product)

For any sets A and B, the *Cartesian product* of A and B, denoted $A \times B$, is the set of all ordered pairs, each of whose first component is an element of A and whose second component is an element of B. Symbolically,

$$A \times B = \{(m,n) \mid m \in A \text{ and } n \in B\}.$$

Example

(i) If $A = \{1,2\}$ and $B = \{3,4\}$, then
$$A \times B = \{(1,3), (1,4), (2,3), (2,4)\}.$$

(ii) If $C = \{a,b\}$ and $D = \{0,5,10\}$, then
$$C \times D = \{(a,0), (a,5), (a,10), (b,0), (b,5), (b,10)\}$$
and
$$D \times C = \{(0,a), (5,a), (10,a), (0,b), (5,b), (10,b)\}.$$

(iii) If $E = \{1,2,3\}$ and $F = \{2,3,4\}$, then
$$E \times F = \{(1,2), (1,3), (1,4), (2,2), (2,3), (2,4), (3,2), (3,3), (3,4)\}.$$

The operation of multiplication of whole numbers can be defined in terms of the Cartesian product of sets, since the number of elements (ordered pairs) in the Cartesian product of two sets is the product of the numbers of elements in those sets.

While the union and intersection of sets can readily be illustrated with Venn diagrams, this is not the case with Cartesian products. Cartesian products can be illustrated, however, with graphs, using a horizontal axis to represent the first set and a vertical axis to represent the second. The ordered pairs are represented by points in the plane, as illustrated in Figures 2.12, 2.13, and 2.14.

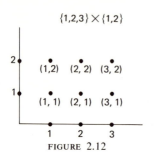

FIGURE 2.12

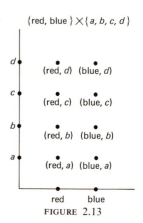

FIGURE 2.13

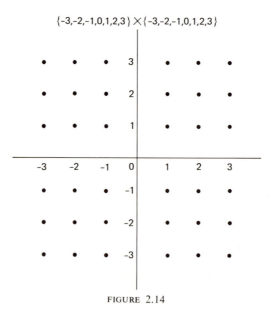

FIGURE 2.14

Exercise Set 2.3
1. Given $A = \{0,1,2\}$, $B = \{1,2,3\}$, $C = \{3,4,5\}$, and $U = \{0,1,2,3,4,5\}$, find
   (a) $A \times B$
   (b) $B \times A$
   (c) $A \times C$
   (d) $B \times C$
   (e) $(A \cup B) \times C$
   (f) $(A \cap B) \times C$
   (g) $(A \times C) \cup (B \times C)$
   (h) $(A \times C) \cap (B \times C)$
   (i) $A \cup (B \times C)$
   (j) $A \cap (B \times C)$
   (k) $A \times \bar{B}$
   (l) $(\overline{A \cup B}) \times C$
   (m) $(B \cap \bar{C}) \times A$
   (n) $(\overline{A \cup C}) \times B$
   (o) $(A \cap C) \times B$
2. Graph the following.
   (a) $\{1,2\} \times \{1,2,3,4\}$
   (b) $\{0,1,2\} \times \{1\}$
   (c) $\{1,2\} \times \{3,4\}$
   (d) $\{-2,-3\} \times \{0,1,2\}$
   (e) $\{a,b,c\} \times \{1,2,3\}$
3. For any set A, find
   (a) $A \times \emptyset$
   (b) $\emptyset \times A$

2.5 PROPERTIES OF SET OPERATIONS

Each operation on sets is defined in terms of logical connectives, for which we have already developed truth tables. Thus, we can use valid argument forms, which have already been established, to obtain properties of set operations. We list these properties as theorems.

theorem 2.1 (the commutative property of intersection)
For any set A and B,

$$A \cap B = B \cap A.$$

This theorem is a direct result of the equivalence

$$p \wedge q \Leftrightarrow q \wedge p.$$

Example
If $A = \{a,b,c\}$ and $B = \{b,c,d\}$, then

$$A \cap B = \{a,b,c\} \cap \{b,c,d\} = \{b,c\}$$

and

$$B \cap A = \{b,c,d\} \cap \{a,b,c\} = \{b,c\}.$$

Therefore,

$$A \cap B = B \cap A.$$

theorem 2.2 (the commutative property of union)
For any sets A and B,

$$A \cup B = B \cup A.$$

This theorem is a direct result of the equivalence

$$p \vee q \Longleftrightarrow q \vee p.$$

Example
If $A = \{a,b,c\}$ and $B = \{b,c,d\}$, then

$$A \cup B = \{a,b,c\} \cup \{b,c,d\} = \{a,b,c,d\}$$

and

$$B \cup A = \{b,c,d\} \cup \{a,b,c\} = \{b,c,d,a\} = \{a,b,c,d\}.$$

Therefore,

$$A \cup B = B \cup A.$$

theorem 2.3 (associative property of intersection)
For any sets A, B, and C,

$$(A \cap B) \cap C = A \cap (B \cap C).$$

This theorem is a direct result of the equivalence

$$(p \wedge q) \wedge r \Longleftrightarrow p \wedge (q \wedge r).$$

Example
   If $A = \{1,2,3,4\}$, $B = \{3,4,5,6\}$, and $C = \{2,4,6,8\}$, then

$$(A \cap B) \cap C = (\{1,2,3,4\} \cap \{3,4,5,6\}) \cap \{2,4,6,8\}$$
$$= \{3,4\} \cap \{2,4,6,8\} = \{4\}$$

and

$$A \cap (B \cap C) = \{1,2,3,4\} \cap (\{3,4,5,6\} \cap \{2,4,6,8\})$$
$$= \{1,2,3,4\} \cap \{4,6\} = \{4\}$$

theorem 2.4 (associative property of union)
   For any sets A, B, and C,

$$(A \cup B) \cup C = A \cup (B \cup C).$$

This theorem is a direct result of the equivalence

$$(p \lor q) \lor r \Longleftrightarrow p \lor (q \lor r).$$

Example
   If $A = \{0,1,2\}$, $B = \{1,2,3\}$, and $C = \{3,4,5\}$, then

$$(A \cup B) \cup C = (\{0,1,2\} \cup \{1,2,3\}) \cup \{3,4,5\}$$
$$= \{0,1,2,3\} \cup \{3,4,5\} = \{0,1,2,3,4,5\}.$$

and

$$A \cup (B \cup C) = \{0,1,2\} \cup (\{1,2,3\} \cup \{3,4,5\})$$
$$= \{0,1,2\} \cup \{1,2,3,4,5\} = \{0,1,2,3,4,5\}.$$

Theorems 2.3 and 2.4 could leave the false impression that the position of parentheses, when associated with set operations, is irrelevant. We have already seen that this is not the case since $A \cup (B \cap C) \neq (A \cup B) \cap C$. Thus, it is important to remember that each associative property deals with one operation only. When an expression involves two operations, however, we often have a *distributive* property that may be used.

theorem 2.5 (distributive property of intersection over union)
For any sets A, B, and C,

$$A \cap (B \cup C) = (A \cap B) \cup (A \cap C).$$

This theorem is a direct result of the equivalence

$$p \wedge (q \vee r) \Leftrightarrow (p \wedge q) \vee (p \wedge r).$$

Example

If $A = \{0,1,2,3,4,5\}$, $B = \{1,3,5,7,9\}$, and $C = \{0,2,4,6,8\}$, then

$$
\begin{aligned}
A \cap (B \cup C) &= \{0,1,2,3,4,5\} \cap \{0,1,2,3,4,5,6,7,8,9\} \\
&= \{0,1,2,3,4,5\}
\end{aligned}
$$

and

$$(A \cap B) \cup (A \cap C) = \{1,3,5\} \cup \{0,2,4\} = \{0,1,2,3,4,5\}.$$

Therefore,

$$A \cap (B \cup C) = (A \cap B) \cup (A \cap C).$$

theorem 2.6 (distributive property of union over intersection)
For any sets A, B, C,

$$A \cup (B \cap C) = (A \cup B) \cap (A \cup C).$$

This theorem is a direct result of the equivalence

$$p \vee (q \wedge r) \Leftrightarrow (p \vee q) \wedge (p \vee r).$$

Example

If $A = \{0,1,2\}$, $B = \{2,3,4\}$, and $C = \{4,5,6\}$, then

$$A \cup (B \cap C) = \{0,1,2\} \cup \{4\} = \{0,1,2,4\}.$$

and

$$
\begin{aligned}
(A \cup B) \cap (A \cup C) &= \{0,1,2,3,4\} \cap \{0,1,2,4,5,6\} \\
&= \{0,1,2,4\}.
\end{aligned}
$$

Therefore,

$$A \cup (B \cap C) = (A \cup B) \cap (A \cup C).$$

We also have distributive properties involving Cartesian product. These also result from a logical connective defined, which corresponds directly to the operation of Cartesian product.

theorem 2.7 (distributive property of Cartesian product over intersection)
For any sets A, B, and C,

$$A \times (B \cap C) = (A \times B) \cap (A \times C).$$

proof
$(m,n) \in A \times (B \cap C)$ is equivalent to stating that $m \in A$ and $n \in B \cap C$. $n \in B \cap C$ is equivalent to ($n \in B$ and $n \in C$). Therefore, ($m \in A$ and $n \in B$) and ($m \in A$ and $n \in C$), which is equivalent to stating that $(m,n) \in (A \times B)$ and $(m,n) \in (A \times C)$. Thus, $(m,n) \in A \times (B \cap C) \Leftrightarrow (m,n) \in (A \times B) \cap (A \times C)$, and we have

$$A \times (B \cap C) = (A \times B) \cap (A \times C).$$

Example
If $A = \{a,b\}$, $B = \{1,2,3\}$, and $C = \{2,3,4\}$, then

$$A \times (B \cap C) = \{a,b\} \times \{2,3\}$$
$$= \{(a,2), (a,3), (b,2), (b,3)\}$$

and

$$(A \times B) \cap (A \times C)$$
$$= \{(a,1), (a,2), (a,3), (b,1), (b,2), (b,3),\}$$
$$\cap \{(a,2), (a,3), (a,4), (b,2), (b,3), (b,4)\}$$
$$= \{(a,2), (a,3), (b,2), (b,3)\}.$$

Therefore,

$$A \times (B \cap C) = (A \times B) \cap (A \times C).$$

theorem 2.8 (distributive property of Cartesian product over union)
For any sets A, B, and C,

$$A \times (B \cup C) = (A \times B) \cup (A \times C)$$

The proof of this theorem is similar to the proof of Theorem 2.7.

Example

If $A = \{0,1\}$, $B = \{1,2\}$, and $C = \{2,3\}$, then

$$
\begin{aligned}
A \times (B \cup C) &= \{0,1\} \times \{1,2,3\} \\
&= \{(0,1), (0,2), (0,3), (1,1), (1,2), (1,3)\}
\end{aligned}
$$

and

$$
\begin{aligned}
(A \times B) \cup (A \times C) &= \{(0,1), (0,2), (1,1), (1,2)\} \\
&\quad \cup \{(0,2), (0,3), (1,2), (1,3)\} \\
&= \{(0,1), (0,2), (1,1), (1,2), \\
&\qquad (0,3), (1,3)\}
\end{aligned}
$$

Therefore,

$$A \times (B \cup C) = (A \times B) \cup (A \times C).$$

The final two theorems, which we give here, are direct results of DeMorgan's laws and normally bear the same name.

theorem 2.9 (DeMorgan's laws)
For any sets A and B,

$$(\overline{A \cap B}) = (\bar{A} \cup \bar{B})$$

and

$$(\overline{A \cup B}) = (\bar{A} \cap \bar{B}).$$

Example

(i) If $A = \{1,2,3,4\}$, $B = \{3,4,5,6\}$, and $U = \{0,1,2, \ldots ,10\}$, then

$$\overline{A \cap B} = \overline{\{3,4\}} = \{0,1,2,5,6,7,8,9,10\},$$

and

$$\bar{A} \cup \bar{B} = \{\overline{1,2,3,4}\} \cup \{\overline{3,4,5,6}\}$$
$$= \{0,5,6,7,8,9,10\} \cup \{0,1,2,7,8,9,10\}$$
$$= \{0,1,2,5,6,7,8,9,10\}.$$

(ii) $\overline{A \cup B} = \{\overline{1,2,3,4,5,6}\} = \{0,7,8,9,10\}$, and

$$\bar{A} \cap \bar{B} = \{\overline{1,2,3,4}\} \cap \{\overline{3,4,5,6}\}$$
$$= \{0,5,6,7,8,9,10\} \cap \{0,1,2,7,8,9,10\}$$
$$= \{0,7,8,9,10\}.$$

Exercise Set 2.4

1. Name the property or properties that justify each of the following statements.
    (a) $\{5,1,3\} \cup \{4,5,6\} = \{4,5,6\} \cup \{5,1,3\}$
    (b) $(\{2,1\} \cap \{1,3\}) \cap \varnothing = \{2,1\} \cap (\{1,3\} \cap \varnothing)$
    (c) $(\{0\} \times \{a,b\}) \cup (\{0\} \times \{1,2\}) = \{0\} \times \{a,b,1,2\}$
    (d) $(\{1,3,5\} \cup \{3,5,7\}) \cap (\{1,3,5\} \cup \{0,7\}) = \{1,3,5\} \cup \{7\}$
    (e) $\{\overline{1,3,5}\} \cap \{\overline{2,4,6}\} = \{\overline{1,2,3,4,5,6}\}$
    (f) $(\{0,5\} \cap \{5,7\}) \cup (\{0,5\} \cap \{0,3\}) = \{0,5\} \cap \{0,3,5,7\}$
    (g) $(B \times A) \cap (B \times \varnothing) = B \times (A \cap \varnothing) = B \times \varnothing = \varnothing$
    (h) $\{\overline{0,2,4,6,8,10}\} \cup \{\overline{2,3,5,7}\} = \{\overline{2}\}$
    (i) $(A \cup B) \cap (\bar{A} \cup B) = (A \cap \bar{A}) \cup B = \varnothing \cup B = B$
    (j) $(\{0,1,5\} \cup \{2,4,6\}) \cap \{1\} = \{1\} \cap (\{0,1,5\} \cup \{2,4,6\})$
2. Use one or more of the properties discussed in this section to simplify each of the following operations. Name the properties used and find the designated sets. $U = \{0,1,2, \ldots ,10\}$.
    (a) $(\{1,2\} \times \{5,6\}) \cap (\{1,2\} \times \{6,7,8,9\})$
    (b) $(\{0,1,2\} \cap \{1,2,3,4\}) \cup (\{0,1,2\} \cap \{0,2,4,6\})$
    (c) $(\{a,b\} \cup \{3,5,7\}) \cap (\{a,b\} \cup \{0,5,10\})$
    (d) $(\{a,b\} \times \{0,2,4\}) \cup (\{a,b\} \times \{1,3,5\})$
    (e) $\{\overline{2,4,6}\} \cup \{\overline{3,6,9}\}$
    (f) $\{\overline{2,3,5,7}\} \cap \{\overline{4,6,8,9,10}\}$
    (g) $(\{0,2\} \cup \{1,3\}) \cap (\{1,3\} \cup \{0,2,4,6\})$
    (h) $(\{5,7,9\} \cap \{0,5,10\}) \cup (\{0,5,10\} \cap \{2,4,6,8,10\})$
    (i) $(\{2,4,6\} \times \{1,3,5,9\}) \cap (\{2,4,6\} \times \{\overline{1,3,5,9}\})$
    (j) $(\{1,3,5,6,7\} \cap \{3,6,9\}) \cap \{0,2,4,8,10\}$
3. The distributive properties stated in this section are often called *left* distributive properties, since the set being distributed is found

on the left. Corresponding right distributive properties can also be proved either using the left distributive properties or in much the same way as the left distributive properties are proved. Prove the following for any sets A, B, and C.

(a) $(A \cup B) \cap C = (A \cap C) \cup (B \cap C)$

(b) $(A \cap B) \cup C = (A \cup C) \cap (B \cup C)$

(c) $(A \cup B) \times C = (A \times C) \cup (B \times C)$

(d) $(A \cap B) \times C = (A \times C) \cap (B \times C)$

4. Does $A \times B = B \times A$? Why?

5. If the number of elements in A is 3, and the number of elements in B is 5,

(a) what is the maximum number of elements that could be in $A \cup B$?

(b) what is the minimum number of elements that could be in $A \cup B$?

(c) what is the maximum number of elements that could be in $A \cap B$?

(d) what is the minimum number of elements that could be in $A \cap B$?

2.6 RELATIONS AND FUNCTIONS

We have discussed the relations "is an element of," "is a subset of," "contains," and "is equal to," but have not discussed relations in general or their properties. A general discussion of relations is not restricted to relations on sets or numbers. There are teacher-student relations, parent-child relations, cost-price relations, cause-effect relations, and many others that are faced daily by each of us. Although such relations may not seem mathematical in nature, we can approach and discuss them and their properties by mathematical methods.

Relations can be perceived and depicted in many ways. One of the most common methods of illustrating relations is through the use of graphs. We have all seen graphs illustrating relationships of inflation to unemployment, air pollution to density of population, or profits to investments. On such graphs, each point or position on the graph is actually an element of the relation and is determined by its respective distances from a horizontal axis and a vertical axis. Thus, each element can be named by an ordered pair, and a relation can be defined as a set of ordered pairs, that is, a subset of a Cartesian product.

definition 2.8 (relation)

A relation R from A to B is a subset of $A \times B$. A relation from A to A is said to be a relation *on* A. If the ordered pair (a,b) is in R, that is, $(a,b) \in R$, then we say that a is related to b, denoted $a \, R \, b$. The set of all first components of R is called the *domain* of R, and the set of all second components of R is called the *range* of R.

Example

If $A = \{0,1,2,3, \ldots ,10\}$, then the relation "is twice as much as" on A is

"is twice as much as" $= \{(0,0), (2,1), (4,2), (6,3), (8,4), (10,5)\}$

which means that the first component of each ordered pair "is twice as much as" the second component of the same pair. The domain is the set

$D_R = \{0,2,4,6,8,10\}$, and the range is the set $R_R = \{0,1,2,3,4,5\}$.

Example

If the relation "is a brother of" from the set $A = \{$Bob, Jim, Tom, Sage$\}$ to the set $B = \{$Carol, Sue, Tina$\}$ is denoted by

"is a brother of" $= \{($Bob, Sue$), ($Tom, Carol$), ($Jim, Carol$)\}$

then Bob is a brother of Sue; Tom is a brother of Carol; and Jim is a brother of Carol. Thus, the domain is $\{$Bob, Tom, Jim$\}$, and the range is $\{$Sue, Carol$\}$. This means that Sage is not the brother of anyone in B, and Tina has no brother in A.

Just as operations have properties, so do relations, but they have no properties in common. This is understandable, since operations and relations are entirely different types of concepts. For example, $A \cup B$ is a set, while $A \subset B$ is a statement about sets. A similar comparison can be made between any operation and relation.

definition 2.9 (relation properties)

For any relation R from A to B,

(i) R is *reflexive* if and only if for any $a \in A$, $(a,a) \in R$; that is, $a\ R\ a$.

(ii) R is *symmetric* if and only if for any $(a,b) \in R$, $(b,a) \in R$; that is, if $a\ R\ b$, then $b\ R\ a$.

(iii) R is *transitive* if and only if for any $(a,b) \in R$ and $(b,c) \in R$, $(a,c) \in R$; that is, if $a\ R\ b$ and $b\ R\ c$, then $a\ R\ c$.

Example

Let R be the relation "is a subset of" on A, where A is the set of all subsets of the whole numbers.

(i)   The relation "is a subset" is reflexive for any sets, since every set is a subset of itself.

(ii) The relation "is a subset of" is not symmetric, because $\{1,2\} \subset \{1,2,3\}$, but $\{1,2,3\} \not\subset \{1,2\}$.

(iii) The relation "is a subset of" is transitive, because for any sets A, B, and C, if $A \subset B$ and $B \subset C$, then $A \subset C$.

Example

Let R be the relation "takes a class with" on the set of all students at Pierce College.

(i) R is reflexive, since every student takes a class with himself.

(ii) R is symmetric, because if x takes a class with y, then y takes a class with x.

(iii) R is not transitive if there is even one case in which the following type of situation occurs: x takes math with y, and y takes English with z, while x takes no class with z.

Example

Let R be the relation "is the opposite sex of" on the set of all people on earth.

(i) R is not reflexive, since it is not true that everyone is the opposite sex of himself.

(ii) R is symmetric, because if m is the opposite sex of w, then w is the opposite sex of m.

(iii) R is not transitive, because Bob is the opposite sex of Nancy, and Nancy is the opposite sex of Phil, but Bob is not the opposite sex of Phil.

definition 2.10 (equivalence relation)

> A relation that is reflexive, symmetric, and transitive is called an *equivalence relation*.

Example

  (i)   The relation "is the same age as" is an equivalence relation on the set of all people.

  (ii)   The relation "is the same sex as" is an equivalence relation on the set of all people.

  (iii)   The relation "lives on the same block as" is an equivalence relation on the set of people in a given town.

In addition to reflexive, symmetric, and transitive properties of relations, one important property is that of having only one element of the range paired with each element of the domain. A relation of this type is called a *function*.

definition 2.11 (function)

> A relation f from A to B is called a *function* if and only if $(a,b) \in R$ and $(a,c) \in R$ implies that $b = c$ (that is, there can be no first component paired with distinct second components). Where x is the domain of f, we say that f is a function from x *into* B, denoted $f: x \to B$. If $(x,b) \in f$, we denote the unique element $b \in B$ by $f(x)$. $f(x)$ is called the *image of* x.

Only the concept of a number is more prevalent at all levels of mathematics than the concept of a function. While a third grade child is studying this concept in terms of a "function machine," calculus students are considering whether or not a certain function is continuous, and algebra students are determining whether or not a given relation is a function. As with relations in general, the concept of a function is used in almost all professions. When a psychologist is able to produce a unique behavior pattern from a given stimulus, a mechanic is able to determine the problem from the sound of an engine, or a physician can diagnose a disease from a given set of symptoms, then the concept of a function is being used. Intuitively, this concept of one and only one second component for every first component means that given the first component, the second component can always be uniquely determined as a *function* of that first component; that is, a function implies predictability. This does not mean that different first components must have different second components, however. For example, we have all seen certain sororities serving the function that all girls who join come out exactly the same, regardless of how they went in.

The concept of a function can be thought of as a relationship between each first component x and the corresponding second component $f(x)$, or as a set of points, no two of which lie on the same vertical line, on a graph whose horizontal axis represents the domain and whose vertical axis represents the range of the relation. This concept can also be thought of in terms of a "function machine" as in Figure 2.15, which has a unique output for each input.

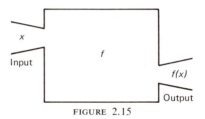

FIGURE 2.15

Example

(i) $\{(1,2)(1,4)(3,5)(2,6)\}$ is not a function because 1 is paired as a first component with both 2 and 4 as second components.

(ii) $\{(1,3)(2,4)(3,5)(4,6)\}$ is a function because no first component is paired with distinct second components.

(iii) $\{(5,1)(6,2)(7,1)(8,2)\}$ is a function because no first component is paired with distinct separate components.

(iv) $\{(x,y)\,|\,x$ is a whole number and $y=2x\}$ is a function.

(v) $\{(x,f(x))\,|\,x=|f(x)|$ and $f(x)$ is an integer$\}$ is not a function because $(3,3)$ and $(3,-3)$ are both elements of the relation.

Exercise Set 2.5

1. Where the following are relations on the set of all people now living, identify the properties (reflexive, symmetric, or transitive) of each relation. If the relation has none of these properties, then so state.

(a) weighs at least as much as

(b) is an ancestor of

(c) is smarter than

(d) is older than

(e) is the same sex as

(f) sits next to

(g) is married to

(h) is a neighbor of

(i) is a sister of

(j) loves

2. Where the following relations are defined on the set of all subsets of W, name the properties of each relation.

(a) $\subset$ (c) $\not\subset$

(b) $=$ (d) $\neq$

3. Determine which of the following relations are functions and which are not.

(a) $\{(1,2)(3,4)(5,6)(7,8)(9,10)\}$

(b) $\{(a,2)(1,1)(2,2)(3,1)(4,2)(5,1)\}$

(c) $\{(5,1)(4,1)(6,1)(-3,1)\}$

(d) $\{(1,5)(1,4)(1,6)(1,-3)\}$

(e) $\{(2,3)(1,4)(3,5)(1,4)(7,2)\}$

4. Is an equivalence relation always a function? Verify your answer.

2.7 SUMMARY

In this chapter, we have defined and discussed concepts and terminology of sets, which underlie many of the topics to be discussed in remaining chapters. We have defined the following terms, along with their symbols.

Contains: $\supset$

Subset: $\subset$

Equality: $=$

Complement: $\bar{A}$

Universe: U

Null Set: $\varnothing$

Union: $\cup$

Intersection: $\cap$

Difference: $A - B$

Cartesian Product: $\times$

Relation: R

Function $f: x \rightarrow B$

We stated the following properties of set operations as theorems. For any sets A, B, and C in universe U,

$(A \cap B) = (B \cap A)$	Commutative property of intersection
$(A \cup B) = (B \cup A)$	Commutative property of union
$(A \cap B) \cap C$ $\quad\quad = A \cap (B \cap C)$	Associative property of intersection
$(A \cup B) \cup C$ $\quad\quad = A \cup (B \cup C)$	Associative property of union
$A \cap (B \cup C)$ $\quad\quad = (A \cap B) \cup (A \cap C)$	Distributive property of intersection over union
$A \cup (B \cap C)$ $\quad\quad = (A \cup B) \cap (A \cup C)$	Distributive property of union over intersection
$A \times (B \cup C)$ $\quad\quad = (A \times B) \cup (A \times C)$	Distributive property of Cartesian product over union
$A \times (B \cap C)$ $\quad\quad = (A \times B) \cap (A \times C)$	Distributive property of Cartesian product over intersection
$\overline{(A \cup B)} = (\bar{A} \cap \bar{B})$ $\overline{(A \cap B)} = (\bar{A} \cup \bar{B})$	DeMorgan's laws

Review Exercise Set 2.6

1. Given that $A = \{0,2,4\}$, $B = \{1,3,5\}$, $C = \{2,3,5\}$, and $U = \{0,1,2,3,4,5\}$, find the following.
   (a) $(A \cup B) \cap C$
   (b) $A \cup (B \cap C)$
   (c) $A \times (B \cap C)$
   (d) $(A \times C) \cap (A \times B)$
   (e) $(\bar{A} \cap \bar{B}) \cap \bar{C}$
   (f) $A \cap (B \times C)$
   (g) $A \cup (B \times C)$
   (h) $(\bar{A} \cup \bar{B}) \cap C$
   (i) $\bar{A} \cup (\bar{B} \cap C)$
   (j) $\bar{A} \cap (\bar{B} \cup C)$
   (k) $A - (B \cup C)$
   (l) $A - \bar{B}$
   (m) $(A - B) \cap \bar{C}$
   (n) $(\bar{A} \cup \bar{B}) - \bar{C}$
   (o) $\overline{(A \cap B)} \cap \overline{(A \cup B)}$.

2. If $A \cap B = \varnothing$, $A \cap C \neq \varnothing$, and $B \cap C \neq \varnothing$, illustrate the following with Venn diagrams.
   (a) $\overline{(A \cup B)} \cap C$
   (b) $\bar{A} \cup (\bar{B} \cap C)$
   (c) $\bar{A} \cap (\bar{B} \cup C)$
   (d) $(A \cup B) \cap \bar{C}$
   (e) $C \cup (\bar{B} \cap A)$

3. State necessary conditions that must hold to make each of the following true.

   (a) $A \cup B = B$ (d) $A \cap \bar{A} = A$
   (b) $A \cap B = B$ (e) $\bar{A} \subset \bar{B}$
   (c) $A \cap \varnothing = A$

4. Identify the properties of each of the following relations.

   (a) "is older than"   (on the set of all people)
   (b) "is at least as tall as"   (on the set of all professional basketball players)
   (c) "is a brother of"   (on the set of all men in this country)
   (d) "is a brother of"   (on the set of all people in this country)
   (e) "is the mother of"   (on the set of all men in this country)
   (f) $\{(1,2)(2,3)(1,3)(1,1)\}$ (on the set of numbers $\{1,2,3\}$)
   (g) $\{(1,1)(2,2)(3,3)(2,3)\}$ (on the set of numbers $\{1,2,3\}$)
   (h) $\{(1,1)\}$ (on the set of numbers $\{1,2,3\}$)
   (i) $\{(1,2), (2,1), (1,1), (2,2)\}$ (on the set of numbers $\{1,2,3\}$)
   (j) $\varnothing$

5. Tell which of the relations in Problem 4 are equivalence relations and which are not.

6. Tell which of the relations in Problem 4 are functions and which are not.

chapter
3
COUNTING AND THE BINOMIAL THEOREM

3.1 INTRODUCTION

Both historically as a species and for each individual person, the first need for the concept of a number has probably been to answer the questions of *how many* objects, *how many* people, *how many* days, or, in general, *how many* elements in a set. A process of determining the number of elements contained in a set is called *counting*. In this chapter, we shall discuss several methods of counting.

3.2 CARDINALITY OF SETS

It is conceivable that the first type of counting that took place was accomplished by some shepherd who, as he turned his flock of sheep out to pasture, placed a pebble in a pouch for each sheep that left. When he wanted to determine whether or not any of the sheep were missing, he merely needed to move a pebble from one pouch to another for each sheep that he observed. If he ran out of sheep before he ran out of pebbles, then he knew sheep were missing. On the other hand, if he ran out of pebbles before he ran out of sheep, then he knew that he had acquired sheep that were not his. If there was one sheep for every pebble and one pebble for each sheep, then he knew he had all of his sheep and only his sheep. This method of counting is forming a *one-to-one correspondence* between the elements of two finite sets. When such a correspondence exists, we say that the two sets have the same *cardinality* or *cardinal number*.

The cardinal number of a set A is the number of elements contained in that set and is denoted by $n(A)$. This number can be determined by setting up a one-to-one correspondence with an ordered subset $\{1,2,3, \ldots ,n\}$ of the positive integers $\{1,2,3, \ldots \}$. The final element of the ordered subset is the cardinal number of the sets. This is essentially the process being used when we "count" the elements of a set making a one-to-one correspondence between the elements and the words "one," "two," "three," and so on.

Example

To determine the cardinality of the set $A = \{a,b,c\}$, we set up the correspondence

$$\{a, b, c\}$$
$$\{1, 2, 3\}.$$

Since the final element of the ordered set $\{1,2,3\}$ is 3, $n(A) = 3$ and $n(\{1,2,3\}) = 3$.

In this way, we can actually define the whole numbers as the cardinal numbers obtained as follows:

$$0 = n(\varnothing)$$
$$1 = n(\{0\})$$
$$2 = n(\{0,1\})$$
$$3 = n(\{0,1,2\})$$
$$\vdots$$
$$m = n(\{0,1,2, \ldots ,m - 1\})$$

with each number being the *successor* of the previous number. We now define addition and multiplication of whole numbers in terms of set operations in such a way that properties of set operations give us properties of whole numbers.

definition 3.1 (addition of whole numbers)
> For any disjoint sets A and B and whole numbers a and b, where $a = n(A)$ and $b = n(B)$, the *sum* of a and b, denoted $a + b$, is defined by

$$a + b = n(A) + n(B) = n(A \cup B).$$

definition 3.2 (multiplication of whole numbers)
> For any sets A and B and whole numbers a and b, where $a = n(A)$ and $b = n(B)$, the *product* of a and b, denoted ab, $(a)b$, $a(b)$, $(a)(b)$, or $a \cdot b$, is defined by

$$ab = n(A) \cdot n(B) = n(A \times B).$$

Example

(i) Let $A = \{a,b,c\}$ and $B = \{2,4,6,8\}$. A and B are disjoint; $n(A) = 3$ and $n(B) = 4$. Therefore, we have

$$3 + 4 = n(A \cup B) = n(\{a,b,c,2,4,6,8\}) = 7$$

and

$$3 \cdot 4 = n(A \times B) = n(\{(a,2), (a,4), (a,6), (a,8), (b,2),$$
$$(b,4), (b,6), (b,8), (c,2), (c,4), (c,6), (c,8)\}) = 12.$$

(ii) If $A = \{1,2,3,4\}$ and $B = \{4,5\}$, then $n(A) = 4$ and $n(B) = 2$, but $4 + 2 \neq n(A \cup B)$ because A and B are *not disjoint* sets; that is, $A \cap B \neq \emptyset$. The definition of multiplication, however, does not require the sets to be disjoint; therefore,

$$4 \cdot 2 = n(A \times B) =$$
$$n(\{(1,4), (2,4), (3,4), (4,4), (1,5), (2,5), (3,5), (4,5)\}) = 8.$$

As direct results of the definitions of addition and multiplication of whole numbers and established properties of set operations, we have the following properties of whole numbers.

properties of whole numbers

For any whole numbers a, b, and c

1. $a + b \in W$ (Closure property of addition)

2. $ab \in W$ (Closure property of multiplication)

3. $a + b = b + a$ (Commutative property of addition)

4. $ab = ba$ (Commutative property of multiplication)

5. $(a + b) + c = a + (b + c)$ (Associative property of addition)

6. $(ab)c = a(bc)$ (Associative property of multiplication)

7. $a(b + c) = ab + ac$ (Distributive property of multiplication over addition)

8. $a + 0 = 0 + a = a$ (Additive identity property)

9. $a \cdot 1 = 1 \cdot a = a$ (Multiplicative identity property)

Subtraction and division are defined in terms of addition and multiplication, respectively, rather than in terms of set operations. This allows us to replace statements involving these operations by equivalent statements involving addition and multiplication and, therefore, have access to the properties of whole numbers that exist with addition and multiplication only.

definition 3.3 (subtraction of whole numbers)

For any whole numbers a, b, and c, the difference of a and b, denoted $a - b$, is defined by $a - b = c$ if and only if $a = b + c$.

definition 3.4 (division of whole numbers)
For any whole numbers a, b, and c with $c \neq 0$, the quotient of a and b, denoted $a \div b = a/b = c$ if and only if $a = bc$.

Exercise Set 3.1

1. If $A = \{0,1,2,3\}$, $B = \{2,3,4\}$, $C = \{4,5,6,7,8\}$, and $U = \{0,1,2, \ldots ,10\}$, find the following.
   (a) $n(A \cup B)$
   (b) $n(A) + n(B)$
   (c) $n(A) + n(C)$
   (d) $n(A \cup C)$
   (e) $n(A \times C)$
   (f) $n(A - B)$
   (g) $n(A \cup \bar{C})$
   (h) $n(\overline{A \cap C})$
   (i) $n(A \cup (B \times C))$
   (j) $n(A \cap (B \times C))$
   (k) $n(\bar{A} \cap \bar{B})$
   (l) $n(A \times \bar{C})$
   (m) $n((A \cup B) \times (A \cup C))$
   (n) $n(\bar{U})$
   (o) $n(A \cup (B \cap C))$

2. Identify each of the following as true or false.
   (a) $n(A \cup B) = n(A) + n(B) + n(A \cap B)$
   (b) If $n(A \cap B) = \varnothing$, then $n(A \cup B) = n(A) + n(B)$
   (c) $n(\{1,2,3\}) + n(\{3,4,5\}) = n(\{1,2,3,4,5\})$
   (d) $n(\bar{A}) = n(U) - n(A)$
   (e) $n(A) + n(\bar{A}) = n(U)$
   (f) $n(A \cup B \cup C) = n(A) + n(B) + n(C) - n(A \cap B \cap C)$
   (g) If $A \subset B$, then $n(A \cup B) = n(A) + n(\bar{B})$
   (h) $n(A \cup B \cup C) = n(A) + n(B) + n(C)$
       $+ n(A \cap B \cap C) - n(A \cap B) - n(A \cap C) - n(B \cap C)$
   (i) $n(\bar{A} \cap \bar{B}) = n(U) - n(A \cap B)$
   (j) $n(\bar{A} \cap \bar{B}) = n(U) - n(A \cup B)$

3. If A is the set of all students taking mathematics and B is the set of all students taking psychology,
   (a) is the number of students taking mathematics or psychology $n(A \cup B)$?
   (b) are there $n(A) + n(B)$ students taking mathematics or psychology?
   (c) how many students are taking both mathematics and psychology?

4. If $n(A) = 7$, $n(B) = 9$, and $n(A \cap B) = 3$, find
   (a) $n(A \cup B)$
   (b) $n(A) + n(B)$
   (c) $n(A - B)$
   (d) $n(B - A)$

5³

5. There are 53 students taking mathematics or physics (or both), 38 students taking mathematics, and 40 students taking physics.
    (a) How many students are taking both mathematics and physics?
    (b) How many students are taking physics but not mathematics?
    (c) How many students are taking mathematics but not physics?
6. An automobile dealer has 100 automobiles on his lot with the following listed as automobiles with accessories:
    40 have vinyl tops
    58 have air conditioning
    72 have power steering and power brakes
    55 have exactly two of the above accessories
    15 have all three accessories
    15 have none of these accessories
    7 have power steering and power brakes, but neither of the other two accessories
    all automobiles with vinyl tops have at least one other accessory
    8 have air conditioning only.
    (a) How many have power steering and power brakes and one other accessory?
    (b) How many have vinyl tops and one other accessory?
    (c) How many have vinyl tops and air conditioning, but no power steering and brakes?
    (d) How many have vinyl tops, air conditioning, and power steering and power brakes?
    (e) How many have power steering and brakes and air conditioning, but no vinyl tops?

3.3 PERMUTATIONS

Two methods of *counting*, which become very useful in the study of probability, as well as the binomial theorem, sequences, series, and other topics within the field of mathematics, are the concepts of *permutations* and *combinations*. In this section, we discuss the concept of a permutation, which is nothing more than an ordered set, with each ordering of the elements of the set being a different permutation.

definition 3.5 (permutation)

An arrangement of the elements of a set *S* in a specific order is called a *permutation* of the elements.

Example

(i)   The possible permutations of the elements of $\{a,b\}$ are ab and ba.

(ii) The possible permutations of the elements of $\{1,2,3\}$ are 123, 132, 213, 231, 312, *and* 321.

(iii) The possible permutations of the elements of $\{1,2,3,4\}$ are 1234, 1243, 1324, 1342, 1423, 1432, 2134, 2143, 2314, 2341, 2413, 2431, 3214, 3241, 3124, 3142, 3421, 3412, 4231, 4213, 4321, 4312, 4123, and 4132.

The number of permutations on a set containing only 1 element is obviously 1. We note from the previous example that the number of permutations possible on a set of 2 elements is 2, or $2 \cdot 1$; the number of permutations possible on a set of 3 elements is 6, or $3 \cdot 2 \cdot 1$; and the number of possible permutations on a set of 4 elements is 24, or $4 \cdot 3 \cdot 2 \cdot 1$. This product of a positive integer with all previous positive integers is called a *factorial* and is defined as follows.

definition 3.6 (factorial)

For any positive integer n, *n-factorial*, denoted $n!$, is the product of all positive integers that are less than or equal to n. Symbolically, $n! = 1 \cdot 2 \cdot 3 \cdot \cdots \cdot n$. We define $0!$ to be 1. If n is a negative integer, then $n!$ is not defined.

Example

$$0! = 1$$
$$1! = 1$$
$$2! = 2 \cdot 1 = 2$$
$$3! = 3 \cdot 2! = 3 \cdot 2 \cdot 1 = 6$$
$$4! = 4 \cdot 3! = 4 \cdot 3 \cdot 2 \cdot 1 = 24$$
$$5! = 5 \cdot 4! = 5 \cdot 4 \cdot 3 \cdot 2 \cdot 1 = 120$$
$$n! = n(n-1)! = n(n-1) \cdot (n-2) \cdots 1$$

As the previous examples illustrate, there are $n!$ possible permutations on n elements, provided that all n elements are distinct and in each permutation.

theorem 3.1

There are $n!$ permutations of n elements taken n at a time. We denote this by $P(n,n) = n!$.

With Theorem 3.1, it becomes a relatively simple matter to determine the number of permutations possible on a given set, but determining what those permutations are is a somewhat more complicated
procedure. One method that can be used is either to mentally or
physically construct a *tree diagram* and systematically obtain all possible permutations of elements beginning with all possible first elements, all possible second elements, and so on.

Example

We can use the following tree diagram (see Figure 3.1) to obtain all possible permutations on the set $\{a,b,c,d\}$.

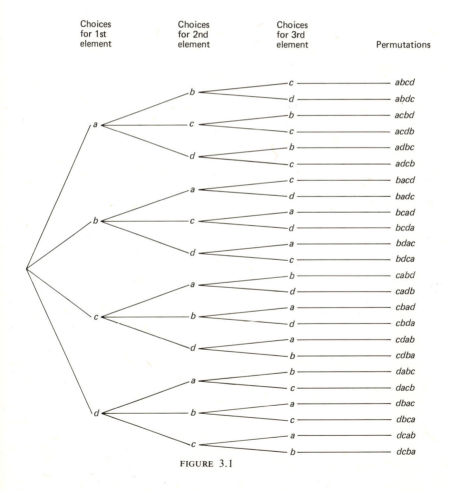

FIGURE 3.1

We sometimes need to know how many permutations can be made on any 4 of 5 elements or any 3 of 6 elements or on any 4 of n elements, where $4 \leq n$. For example, we may be asked to determine all possible three-letter words, using any three of the four letters in the set {n,o,r,t}. Some of these would be "not," "nor," "ton," and "rot." To be certain we have all of them, we must find all possible permutations of the 4 elements taken 3 at a time. The number of such permutations can be determined using the following theorem.

theorem 3.2

The number of permutations of n elements taken r at a time is denoted $P(n,r)$ and is given by the equation

$$P(n,r) = \frac{n!}{(n-r)!}.$$

(*Note:* Theorem 3.1 is merely a special case of Theorem 3.2; that is,

$$P(n,n) = \frac{n!}{(n-n)!} = \frac{n!}{0!} = \frac{n!}{1} = n!$$

Example

(i) The number of permutations of 3 elements taken 2 at a time is

$$P(3,2) = \frac{3!}{(3-2)!} = \frac{3!}{1!} = 3 \cdot 2 \cdot 1 = 6.$$

(ii) The number of permutations of 4 elements taken 2 at a time is

$$P(4,2) = \frac{4!}{(4-2)!} = \frac{4!}{2!} = \frac{4 \cdot 3 \cdot 2!}{2!} = 4 \cdot 3 = 12.$$

(iii) The number of permutations of 6 elements taken 2 at a time is

$$P(6,2) = \frac{6!}{(6-2)!} = \frac{6!}{4!} = \frac{6 \cdot 5 \cdot 4!}{4!} = 6 \cdot 5 = 30.$$

(iv) The number of permutations of 4 elements taken 4 at a time is

$$P(4,4) = \frac{4!}{(4-4)!} = \frac{4!}{0!} = \frac{4!}{1} = 4! = 24.$$

Theorems 3.1 and 3.2 deal with situations in which all of the elements are distinguishable, which is not always the case. For example, assume there are 4 s's, 4 i's, 2 p's, and 1 m in a word and we are trying to determine the number of permutations we can obtain on all 11 letters. We cannot use the formula

$$P(11,11) = 11!$$

because this formula assumes that all 11 letters are distinguishable. The formula we do use is the following theorem.

theorem 3.3

Given n elements, if p are alike, q are alike, r are alike, and so on, the total number of permutations of n elements taken n at a time is

$$P = \frac{n!}{(p!q!r! \ldots)}.$$

Example

There are 11 letters in a word — 4 s's, 4 i's, 2 p's, and 1 m. The number of distinguishable permutations is given by

$$P = \frac{11!}{4!4!2!1!} = \frac{11 \cdot 10 \cdot 9 \cdot 8 \cdot 7 \cdot 6 \cdot 5 \cdot 4!}{4!4 \cdot 3 \cdot 2 \cdot 2 \cdot 1}$$
$$= 11 \cdot 10 \cdot 9 \cdot 7 \cdot 5 = 34{,}650.$$

Example

The number of permutations of 10 balls taken 10 at a time when 5 are black, 3 are red, and 2 are white is

$$P = \frac{10!}{5!3!2!} = \frac{10 \cdot 9 \cdot 8 \cdot 7 \cdot 6 \cdot 5!}{5!3 \cdot 2 \cdot 2 \cdot 1} = 2520.$$

Exercise Set 3.2

1. Compute the following factorials.

    (a) 6!

    (b) 9!

    (c) $5! - 5^2$

    (d) $\dfrac{9!}{8!}$

    (e) $\dfrac{12!}{11!}$

    (f) $\dfrac{29!}{28!}$

    (g) $\dfrac{8!}{42(6!)}$

    (h) $\dfrac{19!}{36(16!)}$

    (i) $\dfrac{1050 \cdot 13!}{16!}$

    (j) $\dfrac{(n+1)!}{(n-1)!}$

2. Find all possible permutations of the following sets.

    (a) {5,6,7}

    (b) {red, black}

    (c) {heads, tails}

    (d) {hearts, spades, diamonds, clubs}

    (e) {red, amber, green}

3. Find the following where all elements are distinguishable.

    (a) $P(5,5)$

    (b) $P(3,3)$

    (c) $P(6,6)$

    (d) $P(6,2)$

    (e) $P(4,3)$

    (f) $P(2,1)$

    (g) $P(10,9)$

    (h) $P(8,7)$

    (i) $P(10,6)$

    (j) $P(8,2)$

    (k) $P(7,3)$

    (l) $P(12,6)$

    (m) $P(18,12)$

    (n) $P(26,5)$

    (o) $P(20,5)$

4. Determine the number of possible permutations of the letters of each of the following words.

    (a) me

    (b) six

    (c) foreign

    (d) work

    (e) sports

    (f) nine

    (g) committee

    (h) eighteen

    (i) three

    (j) school

5. How many different ways may a true-false test be answered if the test contains

    (a) 5 questions

    (b) 7 questions

    (c) 3 questions

    (d) 6 questions

    (e) 10 questions

6. If any permutation of two distinct letters is a word, how many two-letter words can be formed from two distinct letters of each of the following words?
   (a) job
   (b) many
   (c) no
   (d) not
   (e) five
   (f) too
   (g) twelve
   (h) belong
   (i) since
   (j) seven

7. If a coin is tossed repeatedly, how many possible outcomes of the tosses are there if
   (a) the coin is tossed once;
   (b) the coin is tossed twice;
   (c) the coin is tossed three times;
   (d) the coin is tossed four times;
   (e) the coin is tossed five times.

8. If a bucket contains 10 golf balls — 3 white, 4 green, and 3 red — how many different line arrangements are possible?

9. How many different license plates can be made if each has 5 distinct digits?

10. How many different ways can 5 books be arranged between 2 bookends?

11. How many ways can 9 students be seated in 10 desks?

12. A young man is dating 6 girls, but only dates on Friday and Saturday nights. How many weeks will it take to date every possible permutation of two girls the same weekend?

13. How many 2-letter words can be formed from the 26-letter English alphabet if any two letters in any order form a word? How many 3-letter words? How many 4-letter words?

3.4 COMBINATIONS

We have defined permutations to be ordered sets in which each ordering is a distinct permutation from every other ordering. Often, however, we wish to count only those sets with different elements, with no distinction among the different orderings of the elements. When this is the case, and the order in which the elements of a set are considered is not important, then the elements determine a *combination* instead of a permutation.

definition 3.7 (combination)

If the elements of a set are selected without regard to order, then the selection is said to be a *combination*.

Example

While the number of *permutations* of the 4 elements of $\{a,b,c,d\}$ taken 3 at a time is

$$\frac{4!}{(4-3)!} = \frac{4!}{1!} = 4! = 24,$$

the only *combinations* of these 4 elements taken 3 at a time are abc, abd, acd, bcd. Therefore, there are 4 combinations and 24 permutations of the set.

The number of combinations of n elements taken 4 at a time is designated by $C(n,r)$ or $\binom{n}{r}$ and can be determined by the following theorem.

theorem 3.4

For any whole numbers r and n if $r \leq n$, then the number of combinations of n elements taken r at a time is given by the equation

$$C(n,r) = \binom{n}{r} = \frac{P(n,r)}{r!} = \frac{n!}{(n-r)!r!}.$$

Example

(i) The number of combinations of 4 elements taken 3 at a time is

$$\binom{4}{3} = \frac{4!}{(4-3)!3!} = \frac{4!}{1!3!} = \frac{4 \cdot 3!}{1 \cdot 3!} = 4.$$

(ii) The number of combinations of 6 elements taken 3 at a time is

$$\binom{6}{3} = \frac{6!}{(6-3)!3!} = \frac{6!}{3!3!} = \frac{6 \cdot 5 \cdot 4 \cdot 3 \cdot 2 \cdot 1}{3 \cdot 2 \cdot 1 \cdot 3 \cdot 2 \cdot 1} = 20.$$

Example

How many combinations of 4 students can be selected to serve as a refreshment committee for a party involving 10 class officers? We compute the combination of 10 elements taken 4 at a time

$$\binom{10}{4} = \frac{10!}{(10-4)!4!} = \frac{10 \cdot 9 \cdot 8 \cdot 7 \cdot 6!}{6!4 \cdot 3 \cdot 2 \cdot 1} = 210.$$

Example

In how many ways can 3 students be selected for scholarship from a class of 72 students? We compute the combination of 72 students taken 3 at a time; that is,

$$C(72,3) = \binom{72}{3} = \frac{72!}{(72-3)!3!} = \frac{72!}{69!3!}$$

$$= \frac{72 \cdot 71 \cdot 70 \cdot 69!}{69!3 \cdot 2 \cdot 1} = 59,640.$$

The following theorem can be proved immediately from Theorem 3.4.

theorem 3.5

For any whole numbers n and r with $r \leq n$,

$$\binom{n}{r} = \binom{n}{n-r}.$$

Example

$$\binom{12}{2} = \frac{12!}{(12-2)!2!} = \frac{12!}{10!2!} = \frac{12 \cdot 11 \cdot 10!}{10!2 \cdot 1} = 66$$

$$\binom{12}{10} = \frac{12!}{(12-10)!10!} = \frac{12!}{2!10!} = \frac{12 \cdot 11 \cdot 10!}{2 \cdot 1 \cdot 10!} = 66.$$

Exercise Set 3.3

1. Find all possible combinations of 2 elements of each of the following sets.
   (a) $\{a,b,c\}$
   (b) $\{1,2,3,4\}$

(c) {red, green, blue, yellow}

(d) {1,2,3,4,5}

(e) {a,b,c,d,e,f}

2. Compute the following.

(a) $C(3,2)$

(b) $C(5,2)$

(c) $C(7,3)$

(d) $C(7,4)$

(e) $C(15,3)$

(f) $\binom{5}{5}$

(g) $\binom{3}{3}$

(h) $\binom{a}{a}$

(i) $\binom{2}{1}$

(j) $\binom{10}{9}$

(k) $\binom{10}{6}$

(l) $\binom{10}{4}$

(m) $\binom{18}{12}$

(n) $\binom{26}{5}$

(o) $\binom{20}{5}$

3. Use Theorem 3.4 to show that

$$\binom{n}{r} = \binom{n}{n-r}$$

for any whole numbers n and r, where $r \leq n$.

4. In how many different ways can 4 committee members be chosen from 7 class officers?

5. How many subsets containing 4 elements are there to a set containing 12 elements?

6. How many ways can a basketball team be selected from 12 boys? How many if 2 particular boys must be on the team?

7. How many poker hands (5 cards) can be dealt from a standard deck of 52 cards?

8. How many subcommittees of 3 senators each can be selected from 100 senators (any senator may serve on any number of committees, as long as no 3 senators serve together on more than one committee)?

9. A certain school has 6 cheerleaders but only enough funds to send 3 to cheer at the road games. How many road games would have to be scheduled for every combination of 3 cheerleaders to have a chance to travel to exactly one game?

10. In how many different ways can a 5-man court vote to give a majority decision with all 5 men voting?

3.5 THE BINOMIAL THEOREM

A type of expression commonly encountered in algebra and other areas of mathematics is a power of a binomial of the form

$$(x + a)^n$$

where n is a positive integer, and a is any real number. Such an expression is so common that it is very useful to have a systematic method of finding an equivalent expression in expanded form. For small values of n, this can be done by direct multiplication; that is,

$$(x + a)^1 = x + a$$
$$(x + a)^2 = x^2 + 2ax + a^2$$
$$(x + a)^3 = x^3 + 3x^2a + 3xa^2 + a^3$$
$$(x + a)^4 = x^4 + 4x^3a + 6x^2a^2 + 4xa^3 + a^4$$

and

$$(x + a)^5 = x^5 + 5x^4a + 10x^3a^2 + 10x^2a^3 + 5xa^4 + a^5.$$

It is obvious that as n gets larger, $(x + a)^n$ becomes more difficult to expand by direct multiplication. For example, $(x + a)^{100}$ would require an unreasonable amount of time to expand by this method, yet such an expression, or specific terms from such an expression, is often needed. One method of determining the coefficients of each term is by the use of Pascal's triangle, as illustrated in Figure 3.2.

$$
\begin{array}{lccccccccc}
(x + a)^0 & & & & & & 1 & & & \\
(x + a)^1 & & & & & 1 & & 1 & & \\
(x + a)^2 & & & & 1 & & 2 & & 1 & \\
(x + a)^3 & & & 1 & & 3 & & 3 & & 1 \\
(x + a)^4 & & 1 & & 4 & & 6 & & 4 & & 1 \\
(x + a)^5 & 1 & & 5 & & 10 & & 10 & & 5 & & 1 \\
(x + a)^6 & 1 & & 6 & & 15 & & 20 & & 15 & & 6 & & 1 \\
(x + a)^7 & 1 & & 7 & & 21 & & 35 & & 35 & & 21 & & 7 & & 1
\end{array}
$$

. . .

FIGURE 3.2. PASCAL'S TRIANGLE

In this array, the first and last entries in each row are 1, and each other entry is the sum of the entries in the previous row to its left and its right. This method is not always practical, however, because it gives no way of determining the coefficients of $(x + a)^n$ until the coefficients have been determined for $(x + a)^0$, $(x + a)^1$, $(x + a)^2$, $(x + a)^3$, . . . , and $(x + a)^{n-1}$. Thus, we state the following theorem, which is commonly called the *binomial theorem*. The proof of this theorem makes use of the principle of finite induction, which is discussed in Appendix B of this book; so the proof is found in the same appendix.

theorem 3.6 (the binomial theorem)

For any positive integer n and real numbers a and b,

$$(a + b)^n = a^n + \binom{n}{1} (a^{n-1}b) + \binom{n}{2} (a^{n-2}b^2)$$

$$+ \binom{n}{3} (a^{n-3}b^3) + \cdots + \binom{n}{n-1} (ab^{n-1}) + b^n$$

in which the coefficient of $a^{(n-r)}b^r$ is $\binom{n}{r}$ for any $0 \leqslant r \leqslant n$.

Example

(i) $(x + a)^5$

$$= x^5 + \binom{5}{1} x^4a + \binom{5}{2} x^3a^2 + \binom{5}{3} x^2a^3 + \binom{5}{4} xa^4 + a^5$$

$$= x^5 + \frac{5!}{4!} x^4a + \frac{5!}{3!2!} x^3a^2 + \frac{5!}{2!3!} x^2a^3 + \frac{5!}{4!} xa^4 + a^5$$

$$= x^5 + 5x^4a + 10x^3a^2 + 10x^2a^3 + 5xa^4 + a^5.$$

(ii) $(3 + a)^4 = 3^4 + \binom{4}{1} 3^3a + \binom{4}{2} 3^2a^2 + \binom{4}{3} 3a^3 + a^4$

$$= 81 + \frac{4!}{3!} 27a + \frac{4!}{2!2!} 9a^2 + \frac{4!}{3!} 3a^3 + a^4$$

$$= 81 + 108a + 54a^2 + 12a^3 + a^4.$$

Example

Find the 9th term of $(x + a)^{15}$.

The rth term of $(x + a)^n$ is

$$\binom{n}{r-1} x^{n-(r-1)} a^{r-1},$$

so the 9th term of $(x + a)^{15}$ is

$$\binom{15}{8} x^7 a^8 = \frac{15!}{7!8!} x^7 a^8 = 6435 x^7 a^8.$$

Example

Find the coefficient of $x^7 y^9$ in the expansion of $(x + y)^{16}$. The coefficient of $a^{n-r} b^r$ in the expansion of $(a + b)^n$ is $\binom{n}{r}$, so the coefficient of $x^7 y^9$ is

$$\binom{16}{9} = \frac{16!}{7!9!} = 11{,}440.$$

Exercise Set 3.4

1. Use Pascal's triangle to expand the following.
    (a) $(x - 5b)^3$
    (b) $(2x + 3)^4$
    (c) $(2x + 3)^3$
    (d) $(a - 3b)^3$
    (e) $(3 + 2a)^4$

2. Use the binomial theorem to expand the following.
    (a) $(a + 2b)^3$
    (b) $(3a - b)^4$
    (c) $(2a + 3b)^4$
    (d) $(17a - 3)^3$
    (e) $(2a - 7b)^5$

3. Find the coefficient of each of the following.
    (a) $x^3 y^4$ in the expansion of $(x + y)^7$
    (b) $x^5 y^3$ in the expansion of $(x + y)^8$
    (c) the 5th term of $(a + b)^7$
    (d) the 9th term of $(a - b)^{12}$
    (e) the 8th term of $(x + y)^{14}$
    (f) the term of $(x^2 - 3y^3)^{11}$ that contains a factor of x^{14}
    (g) the term of $(x^3 + y^2)^{10}$ that contains a factor of y^6
    (h) the 4th term of $(x + 2)^6$
    (i) $x^7 y^2$ in the expansion of $(x + y^2)^8$
    (j) $x^{12} y^4$ in the expansion of $(x^3 + y)^8$

4. Find the following terms.
    (a) the 8th term of $(x + y)^{14}$
    (b) the 5th term of $(x + 2)^7$
    (c) the first 3 terms of $(x + 3)^{14}$
    (d) the first 3 terms of $(x + 3)^9$
    (e) the first 3 terms of $(10 - 4)^5$

3.6 PARTITIONS AND THE MULTINOMIAL THEOREM

In determining combinations of elements of a set, we often wish to consider only combinations that are themselves disjoint sets. For example, if we wish to have a class separate into discussion groups that will all meet simultaneously, then any two discussion groups are disjoint sets of students. We may also wish to have everyone involved in one of these discussion groups. If this is the case, then such a division or separation of the class becomes a *partition*.

definition 3.8 (partition)

For every set S, a collection of nonempty subsets S_1, S_2, S_3, . . . , S_n of S is called a partition of S, provided that the following two conditions hold:

1) For any S_h and S_k in the collection of sets S_1, S_2, S_3, . . . , S_n, $S_h \cap S_k = \emptyset$; that is, all sets in the partition are disjoint sets.
2) $S_1 \cup S_2 \cup S_3 \cup \cdots \cup S_n = S$.

Example

(i) Any two of the sets $A = \{1\}$, $B = \{4,6,8,9,10\}$, $C = \{0\}$, and $D = \{2,3,5,7\}$ are disjoint, and the union of all of the sets is the set $S = \{0,1,2,3,4,5,6,7,8,9,10\}$. Therefore, the sets A, B, C, and D form a partition on the set S.

(ii) The sets $A = \{0,2,4,6,8,10\}$, $B = \{2,3,5,7\}$, and $C = \{1,9\}$ do not determine a partition on any set since $A \cap B \neq \emptyset$.

(iii) The sets $A = \{4,6,8,9,10\}$, $B = \{2,3,5,7\}$, and $C \neq \{0\}$ do not form a partition of the set $S = \{0,1,2, . . . ,10\}$ because $A \cup B \cup C \neq S$.

Now let us carry the concept of a partition one step further. Suppose in the example of separating the class into discussion groups that each group is to discuss a different topic from that of any other group, or have a different number of members, or is in some other way distinguishable. If this is the case, then the order in which each group is assigned is distinguishable from any other order of assignments. Thus, we have what we call an *ordered partition*.

definition 3.9 (ordered partition)

Let S be a set containing n elements. Any partition of S consisting of r sets with n_1 elements in the first set, n_2 elements in the second set, n_3 elements in the third set, and so on, to n_r elements in the rth set is called an *ordered partition*. The number of such ordered partitions possible is denoted by the symbol

$$\binom{n}{n_1,n_2,n_3,\ \ldots\ ,n^4}.$$

The following theorem is an immediate result of the definition of addition of whole numbers and the definition of a partition.

theorem 3.7

For any partition S_1, S_2, S_3, . . . , S_r on S, where $n_1 = n(S_1)$, $n_2 = n(S_2)$, $n_3 = n(S_3)$, . . . , $n_r = n(S_r)$, and $n = n(S)$, we have $n_1 + n_2 + n_3 + \cdots + n_r = n$.

With this theorem, we are now able to state the following theorem, which contains a formula that can be used to compute the number of possible ordered partitions of a set.

theorem 3.8

For any set S of n elements,

$$\binom{n}{n_1,n_2,n_3,\ \ldots\ ,n_r} = \frac{n!}{n_1!n_2!n_3!\ \ldots\ n_r!}.$$

Example

If a patrol of 11 scouts is partitioned into groups of 3 boys to set up tents, 2 boys to gather firewood, 4 boys to clear the camp ground, and 2 boys to cook, how many different ways are there of making the assignments?

$$\binom{n}{n_1,n_2,n_3, \ldots ,n_r} = \frac{n}{n_1!n_2!n_3! \ldots n_r!}.$$

So

$$\binom{11}{3,2,4,2} = \frac{11!}{3!2!4!2!} = 69,300.$$

Example

If 4 roommates decide they will separate into groups of 2 with one group to do the cleaning and the other the cooking, how many different ways can the assignments be made?

$$\binom{4}{2,2} = \frac{4!}{2!2!} = \frac{4 \cdot 3 \cdot 2}{2 \cdot 2} = 6.$$

Example

If there are only 2 seats left in an English class, 5 in a mathematics class, and 4 in a physics class, in how many ways can 11 students register to fill these vacancies?

$$\binom{11}{2,5,4} = \frac{11!}{2!5!4!} = 6930.$$

One valuable application of the formula for computing the number of ordered partitions of a set is in the *multinomial theorem*. This theorem is similar to the binomial theorem in that it provides us with a method of immediately expanding an algebraic expression of the form $(x_1 + x_2 + x_3 + \cdots + x_r)^n$ for any positive integer n, and also a method of determining any particular term or terms of the expansion.

theorem 3.9 (the multinomial theorem)

The expansion of $(x_1 + x_2 + x_3 + \cdots + x_r)^n$ is found by adding all terms of the form

$$\binom{n}{n_1, n_2, n_3, \ldots, n_r} x_1^{n_1} x_2^{n_2} x_3^{n_3} \cdots x_r^{n_r}$$

where n_1, n_2, n_3, $\ldots$, n_r are nonnegative integers such that $n_1 + n_2 + n_3 + \cdots + n_r = n$.

Example

The expansion of $(a + b + c)^3$ is

$$\binom{3}{3,0,0} a^3 + \binom{3}{2,1,0} a^2 b + \binom{3}{2,0,1} a^2 c + \binom{3}{1,2,0} ab^2$$
$$+ \binom{3}{0,2,1} b^2 c + \binom{3}{1,1,1} abc + \binom{3}{0,3,0} b^3 + \binom{3}{1,0,2} ac^2$$
$$+ \binom{3}{0,1,2} bc^2 + \binom{3}{0,0,3} c^3 = a^3 + 3a^2 b + 3a^2 c + 3ab^2 + 3b^2 c$$
$$+ 6abc + b^3 + 3ac^2 + 3bc^2 + c^3.$$

Example

The coefficient of $x^4 y z^3 w^2$ in the expansion of $(x + y + z + w)^{10}$ is

$$\binom{10}{4,1,3,2} = \frac{10!}{4!1!3!2!} = 12,600.$$

Exercise Set 3.5

1. Identify each of the following as a partition or not a partition of the set $S = \{0,1,2, \ldots ,20\}$.
   (a) $\{0,1,2,3\}$, $\{5,7,9,11\}$, $\{4,6,8,10,12\}$, $\{13,15,17,19\}$, $\{14,16,18,20\}$
   (b) $\{0,2,4, \ldots ,20\}$, $\{1,3,5, \ldots ,20\}$
   (c) $\{0,2,4, \ldots ,20\}$, $\{0,1,3,5,7,9,11,13,15,17,19\}$
   (d) $\{1,3,5,7,9\}$, $\{11,13,15,17,19\}$, $\{2,4,6, \ldots ,20\}$
   (e) $\{0,2,4, \ldots ,20\}$, $\{3,6,9, \ldots ,18\}$, $\{5,10,15,20\}$, $\{7,11,13,17,19\}$

2. Compute the following.

(a) $\binom{5}{3,2}$

(i) $\binom{9}{4,3,2}$

(b) $\binom{7}{4,3}$

(j) $\binom{12}{5,2,3,2}$

(c) $\binom{8}{4,4}$

(k) $\binom{8}{2,2,2,2}$

(d) $\binom{8}{5,3}$

(l) $\binom{9}{6,2,1}$

(e) $\binom{8}{6,2}$

(m) $\binom{9}{5,0,4}$

(f) $\binom{5}{2,2,1}$

(n) $\binom{16}{12,2,2}$

(g) $\binom{7}{3,2,2}$

(o) $\binom{20}{14,2,2,2}$

(h) $\binom{8}{4,2,2}$

3. Find the coefficient of each of the following.
   (a) x^2yz in the expansion of $(x+y+z)^4$
   (b) a^3b^3c in the expansion $(a+b+c)^7$
   (c) a^5bc^3 in the expansion $(a+b+c)^9$
   (d) $x^3y^4zw^5$ in the expansion of $(x+y+z+w)^{13}$
   (e) $xy^2z^3wt^4$ in the expansion of $(x+y+z+w+t)^{11}$

4. If 5 fraternity brothers decide to date 2 co-eds over the next week, with 3 dating the blonde and 2 dating the brunette, how many different ways could the assignments of man to woman be made (disregarding the days the dates are to take place)?

5. In how many ways can 9 students be assigned to 3 dormitories with 3 being assigned to each dormitory?

6. In how many ways can 12 boys be assigned to 2 basketball teams with 1 substitute to each team? Consider the position of substitute separate from that of team member and disregard different positions on the team.

7. In how many ways can 12 toys be divided among 3 children if the oldest one gets 6, the second gets 4, and the third gets 2?

8. In how many ways can a football team end a 14-game season with 7 wins, 5 losses, and 2 ties?

9. In how many ways can 25 cards be dealt, with 5 cards going to each of 5 players?
10. Expand the following multinomials.

(a) $(a + b + c)^4$ (d) $(a + b + c + d)^3$

(b) $(a + b + c)^5$ (e) $(a + b + c + d + e)^3$

(c) $(a + b + c + d)^2$

3.7 SUMMARY

In this chapter, we have defined counting as determining the number of elements in a set and have discussed four methods of counting. The first method, that of demonstrating a one-to-one correspondence with the numbers 1, 2, 3, . . . , n for some number n, is the method most commonly used by people in virtually all walks of life.

The second method is that of determining the number of permutations possible on the elements of a set, with a permutation defined as an ordering of the elements of a set. The formula used is

$$P(n,r) = \frac{n!}{(n-r)!}$$

where n represents the number of elements in the set, and r represents the number of elements in each permutation; that is, $P(n,r)$ is the number of possible permutations on n elements taken r at a time.

The third method of counting, which was considered, is determining the number of possible *combinations* of elements of a set. A combination differs from a permutation in that we ignore the order of the elements in a combination. The formula used for computing the number of combinations of r elements that can be found in a set of n elements is

$$C(n,r) = \binom{n}{r} = \frac{P(n,r)}{r!} = \frac{n!}{(n-r)!\,r!}.$$

The final method of counting discussed was that of determining the number of *ordered partitions* possible on a set. The formula used when n is the number of elements in the set and n_1, n_2, n_3, . . . , n_r represents the number of elements in each of the sets of the partition is

$$\binom{n}{n_1, n_2, n_3, \ \ldots \ , n_r} = \frac{n!}{n_1! n_2! n_3! \ \ldots \ n_r!}.$$

The terms defined in this chapter are the following:

Cardinality
Addition of Whole Numbers
Multiplication of Whole Numbers
Subtraction of Whole Numbers
Division of Whole Numbers
Permutation
Factorial
Combination
Partition
Ordered Partition

Review Exercise Set 3.6

1. If $A = \{1,3,5,7,9\}$, $B = \{0,2,4,6,8\}$, $C = \{0,1,2,3,5,7\}$, and $U = \{0,1,2, \ \ldots \ ,10\}$, find the following.
   (a) $n(A \cup B)$
   (b) $n(A \cup C)$
   (c) $n(B \cap C)$
   (d) $n(A \cup (B \cap C))$
   (e) $n((A \cap B) \cap C)$
   (f) $n(A \times B)$
   (g) $n(A \times (B \cap C))$
   (h) $n(\bar{A} \cap C)$
   (i) $n(\overline{A \cap B})$
   (j) $n(\bar{A} \cup (B \cap C))$

2. Compute the following.
   (a) $P(3,2)$
   (b) $P(7,5)$
   (c) $P(9,2)$
   (d) $P(12,5)$
   (e) $P(16,3)$
   (f) $P(28,2)$
   (g) $P(32,4)$
   (h) $P(17,3)$
   (i) $P(21,2)$
   (j) $P(17,15)$

3. Compute the following.
   (a) $C(3,2)$
   (b) $C(7,5)$
   (c) $C(9,2)$
   (d) $C(12,5)$

(e) $C(16,3)$

(f) $\binom{28}{2}$

(g) $\binom{32}{4}$

(h) $\binom{17}{3}$

(i) $\binom{21}{20}$

(j) $\binom{17}{15}$

4. Compute the following.

(a) $\binom{12}{6,3,3}$

(b) $\binom{9}{5,4}$

(c) $\binom{10}{7,3}$

(d) $\binom{16}{4,8,4}$

(e) $\binom{15}{5,5,5}$

(f) $\binom{11}{7,3,1}$

(g) $\binom{9}{5,0,4}$

(h) $\binom{12}{2,1,3,6}$

(i) $\binom{20}{15,2,2,1}$

(j) $\binom{25}{15,5,2,3}$

5. Prove that for any positive integers n and k such that $k \leqslant n - 1$,

$$\binom{n+1}{k+1} = \binom{n}{k} + \binom{n}{k+1}.$$

6. Using the binomial and multinomial theorems, find
   (a) the 3rd term of the expansion of $(a + b)^{12}$;
   (b) the 6th term of the expansion of $(5 - a)^9$;
   (c) the coefficient of a^4b^3 in the expansion of $(a + b)^7$;
   (d) the coefficient of x^8y^3 in the expansion of $(x + y)^{11}$;
   (e) the coefficient of the term with x^9y^3 as a factor in the expansion of $(x + y + z)^{12}$;
   (f) the term with xy^3 as a factor in the expansion of $(x + y + z)^7$;
   (g) the coefficient of $xy^3z^4w^2$ in the expansion of $(x + y + z + w)^{10}$;
   (h) the coefficient of y^5z in the expansion of $(x + y + z + w)^6$.
7. Identify each of the following as true or false.
   (a) $(4!) \cdot (3!) = 12!$
   (b) $\dfrac{4 \cdot 3!}{4!} = 1$

(c) $\dfrac{(6!) \cdot (4!)}{5!} = 144$

(d) $\dfrac{3!}{6!} = \dfrac{1}{2!}$

(e) $\dfrac{6!5!}{7!} = \dfrac{5!(6-1)}{5! \cdot 6 \cdot 7}$

8. You are an employer and have set certain criteria that your employee(s) must meet to obtain a raise or a promotion. It is your desire that 70 percent of your employees will meet these criteria after the first year. You take a survey and find that out of 96 employees, 20 will receive a raise, 15 a promotion, and 10 both. How many employees get neither a raise nor a promotion?

9. A salesman must travel to Big Rock City from his office once each week. In order to get to Big Rock, he must pass through Small Town. There are three ways to get to Small Town and two routes from Small Town to Big Rock. If he chooses to use a different route each trip and returns by the same route, how long will it take him to use all possible routes from his office to Big Rock?

10. How many different team arrangements can be made for a doubles tennis match if each team must be composed of 1 boy and 1 girl from a class of 5 girls and 4 boys?

11. Suppose you wrote a textbook consisting of 7 chapters, you decide that 2 of the chapters must make the first two chapters of the book. How many different chapter arrangements may you pick from to get a chapter arrangement?

12. The Peterson Corp., Ltd., runs regular checks of common stocks and rates them as one of the following.

NR = No Risk   R = Risky

VLR = Very Little Risk   VR = Very Risky

SR = Slight Risk   NC = No Chance

(a) How many ways can two different stocks be rated?

(b) How many ways can two different stocks be rated if all we know is that their ratings are not the same?

(c) How many ways can three different stocks be rated if they rate at least SR(Slight Risk)?

13. A sales manager for Macy's Dept. Store has been instructed that the primary colors (red, yellow, and blue) attract the greatest attention to window displays. The sales manager has 8 dresses, consisting of 2 reds, 3 blues, 1 yellow, and 2 greens. If he mixes

the dresses in any order, how many arrangements does he have from which to choose? If he chooses to use only the red, yellow, and blue, how many does he have?

14. Suppose you are a small businessman and must restock your shelves. You have five products that you must buy in an order of priority due to lack of funds. If you desire to evaluate every possible arrangement of the products in order to decide on the proper order of product/priority arrangements, how many arrangements do you have from which to choose? How many arrangements do you have if you decide that two particular products must go 1st and 2nd on the list?

15. You are an employer that hires college students for summer help. You have 15 students (boys and girls) and have jobs that consist of 4 employees for yard cleanup, 3 clerical, 2 building janitorial, 3 shelf restocking, and 3 for door-to-door salesmen or women. In how many ways can you divide the students up? If you evaluate an arrangement each minute, how long will it take?

16. A high school science fair has 50 entries, of which 6 finalists are to be chosen. The 6 finalists are to be questioned and judged on their overall understanding of their project, and a winner will be chosen. How many ways can the 6 finalists and winner be chosen?

17. Suppose you are a recreation supervisor to a youth summer camp and have split the camp into 12 team leagues with each team consisting of 4 men. If each team is to play every other team exactly twice, how many league games will be played?

4
PROBABILITY

4.1 INTRODUCTION

In the uncertain world in which we live, each of us is constantly faced with questions such as "What are the chances that I will graduate from college?", "What are my chances of being drafted?", "What is the probability that my marriage will be successful?", "What are the chances of succeeding in a particular business?", "What is the probability of rain today?", "What is the probability that this investment will pay off?", and "What are my chances of living long enough to pay my home off?". All of these questions deal with chance, or *probability,* the study of which we call *probability* theory. Through a study of probability theory, we are able to mathematically compute levels of confidence, that we may have in predictions of certain outcomes of uncertainties of the physical world in which we live.

There are two types of probability that concern most people. One type, called _personalistic probability,_ deals with predictions of specific events that cannot be repeated under exactly the same circumstances, such as the answers to the questions in the previous paragraph. This type of probability is influenced by personal beliefs, individual abilities, and numerous other variables that cannot be precisely measured. Another type of probability, called _objective probability,_ deals with what may be expected to happen in repeated trials under precisely the same conditions. This is the type of probability involved with dice, cards, and other games of chance. In fact, much probability theory was developed for the express purpose of aiding in games of chance. For example, the seventeenth-century French gambler, the Chevalier de Méré, after losing much money in games of chance, asked the mathematician Blaise Pascal (1623–1662) how to determine the probability each player has of winning a dice game at any given stage. Pascal and Pierre de Fermat (1601–1665) discussed the problem, and the solution was found. Likewise, other mathematicians became involved in the study of probability, and the topic gradually emerged in publications of many prominent scientists.

While the discussion of probability in this book is restricted to "objective probability," we shall also discuss some topics affecting "personalistic probability" under the title of _statistics._

Before we can discuss probability theory, we must first define the basic terminology to be used. First, we consider the process by which an observation is made to be an _experiment,_ and the result of the experiment is called an _outcome._ Every possible outcome, no two of which may be outcomes at the same time, is called a _sample point._ A nonempty set of all sample points of an experiment is called a _sample space,_ and any subset of a sample space is an _event._

Example

One type of experiment familiar to most people is what we call a spinner. A spinner consists of a needle fixed to a circular piece of cardboard, so that it will spin in such a way that it is equally likely to stop at any one place as at any other. The experiment is the spinning of the needle, and the outcome is the position in which the needle stops. If the needle stops on a line between two positions, we assign the nearest outcome in a clockwise direction to be the outcome of the experiment. Consider the spinner illustrated in Figure 4.1, with the sample space {1,2,3,4,5,6,7,8} and 1 as the outcome shown. All subsets of the sample space are events. For example, some of the events that may be considered are those in which the needle stops on an even-numbered space, {2,4,6,8}, an odd-numbered space, {1,3,5,7}, a space numbered less than 4, {1,2,3}, a space numbered greater than 6, {7,8}, or a space between 3 and 7, {4,5,6}. In fact, each singleton set containing only one sample point is an event.

FIGURE 4.1

Example

Consider the spinner in Figure 4.2. Let the experiment consist of spinning the needle and observing whether it stops on a region named by an integer that is less than 4 or an integer that is greater than four. The set of outcomes being considered is {less than 4, greater than 4} with only two elements. Is this set a sample space of the experiment? Since the needle could stop on 4, the set {less than 4, greater than 4} does not include all possible outcomes, so it is *not* a sample space.

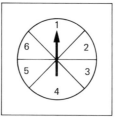

FIGURE 4.2

Another type of experiment, which is familiar to most people, is coin tossing.

Example

Consider the experiment in which two coins are tossed. The possible outcomes are a head on both coins, HH, a tail on both coins, TT, a head on the first coin and a tail on the second coin, HT, and a tail on the first coin and a head on the second, TH. Thus, the sample space has four sample points, {HH,TT,HT,TH}. In addition to the individual sample points, some of the events that may be considered are "heads on one coin and tails on the other" {HT,TH}, "both coins the same" {HH,TT}, "heads on the first coin" {HH,HT}, "tails on the first coin" {TT,TH}, and "heads on the second coin" {HH,TH}.

Example

Consider the experiment of tossing three coins and observing whether the outcome of each is a head or a tail. The sample space is

$$S = \{HHH,HHT,HTH,HTT,THH,THT,TTH,TTT\}.$$

Some of the events associated with S are
(i)   all three coins showing the same {HHH,TTT}
(ii) two coins alike and the other different
       {HHT,HTT,TTH,THT,THH,HTH}
(iii) two heads and one tail   {HHT,HTH,THH}
(iv) all coins heads   {HHH}
(v)   no heads and no tails $\varnothing$
(vi) all coins tails   {TTT}.

When an event contains only one sample point, as in Parts (iv) and (vi) of the previous example, the event is called a *simple event*. When it contains more than one point, as in Parts (i), (ii), and (iii), it is called a *compound event*. Thus, every event is either a simple event, a compound event, or the empty set.

Note that not all of the events associated with a sample space are disjoint; that is, they may have sample points in common. Those that are disjoint are said to be *mutually exclusive* events. In the previous example, events (i) and (ii) are mutually exclusive, and events (iv) and (vi) are mutually exclusive, but events (ii) and (iii) are not mutually exclusive. Mutually exclusive events are events that cannot occur simultaneously.

Note that Parts (ii), (iii), and (v) of the previous example employ the word "and" in the description of events, thus are *conjunctions* of two or more events. In like manner, we can use the word "or" to define events as *disjunctions* of two or more events. Since events are sets, conjunctions and disjunctions of events are intersections and unions, respectively, of sets. For example, let A and B be events in sample space S. By definition of intersection of sets, $A \cap B$ is the set of all sample points common to A *and* B; that is, A occurs *and* B occurs. By definition of union of sets, $A \cup B$ is the set of all sample points found in A *or* B; that is, A occurs *or* B occurs. Thus, the symbols $\cap$ and $\cup$ may be read "and" and "or," respectively, with reference to probability. For example, if A is the event that the needle of a spinner stops on an even integer and B is the event that the needle stops on an integer less than 5, then $A \cap B$ is the event that the needle stops on an integer that is even *and* less than 5, while $A \cup B$ is the event that the needle stops on an integer that is even *or* less than 5.

Example

Consider the experiment of tossing the three coins and let

A be the event in which there is a head on the first coin;
B be the event in which at least two coins are the same; and
C be the event in which at least one coin shows a tail.

(i) $A \cap B = \{HHT, HTT, HTH, HHH\}$
(ii) $A \cup B = \{HHH, HHT, HTT, HTH, THH, THT, TTH, TTT\}$
(iii) $A \cap C = \{HHT, HTT, HTH\}$
(iv) $(A \cap B) \cap C = \{HHT, HTT, HTH\}$
(v) $A \cup (B \cap C)$
      $= \{HHH, HHT, HTH, HTT\}$
      $\cup \{HHT, HTH, THH, HTT, THT, TTH, TTT\}$
      $= \{HHH, HHT, HTH, HTT, THH, THT, TTH, TTT\}$
(vi) $A \cap (B \cup C)$
      $= \{HHH, HHT, HTH, HTT\}$
      $\cap \{HHH, HHT, HTH, HTT, THH, THT, TTH, TTT\}$
      $= \{HHH, HHT, HTH, HTT\}$.

When discussing sets, the term "complement" is often used with the complement of a set A, denoted $\bar{A}$, being the set of all elements of the universe that do not belong to A. Likewise, in discussing probability, we often refer to *complementary events* as events having this same relationship; that is, if A is an event, then $\bar{A}$ is the set of all sample points of the sample space that do *not belong* to A. Another way of stating the relationship between A and $\bar{A}$ is to say that $\bar{A}$ is the event that A *does not occur*.

Example

(i)   Consider the spinner in the following. If A is the event that the needle stops on a space numbered less than 4, then $A = \{1,2,3\}$ and $\bar{A} = \{4,5,6,7,8\}$.

(ii) Consider the experiment of tossing two coins. If A is the event that both coins come up the same, then $A = \{HH,TT\}$ and $\bar{A} = \{HT,TH\}$.

Exercise Set 4.1

1.   Describe sample spaces for each of the following experiments.

(a)

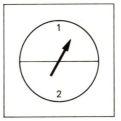

(b)

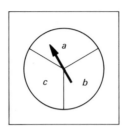

(c)

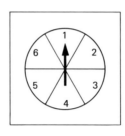

(d) Tossing two coins.

(e) Tossing three coins.

(f) Rolling one six-sided dice.

(g) Choosing a number from the whole numbers 1 to 10 inclusive.

(h) Tossing one coin and one six-sided dice.

2. Consider the experiment of tossing one coin and a regular six-sided dice.

    (a) Given that A is the event that the coin comes up heads and the dice comes up even, that is, $A = \{H2,H4,H6\}$, find $\bar{A}$.

    (b) Given that B is the event that the coin comes up heads and the dice is less than 4, find $\bar{B}$.

    (c) Given that C is the event that the coin comes up tails and the dice is odd, find $\bar{C}$.

    (d) Given that D is the event that the coin comes up tails and the dice is greater than 4, find $\bar{D}$.

3. With A, B, C, and D as defined in Problem 2, find

    (a) $A \cup B$ (f) $C \cup D$

    (b) $A \cap B$ (g) $C \cap D$

    (c) $A \cup C$ (h) $A \cap D$

    (d) $B \cup C$ (i) $B \cup D$

    (e) $A \cup D$ (j) $\bar{B} \cup \bar{D}$

(k) $A \cap \bar{D}$	(p) $\bar{B} \cap C$
(l) $\bar{B} \cap D$	(q) $\bar{A} \cap C$
(m) $\bar{A} \cap D$	(r) $\bar{A} \cap \bar{B}$
(n) $A \cap \bar{D}$	(s) $\bar{C} \cap \bar{D}$
(o) $B \cap \bar{C}$	(t) $\overline{C \cup D}$

4. A political poll is made in which a sample of people are asked to list the political party that they identify as being closest to their own philosophy. If those responses that listed the Republican party are tallied on one list and those that listed the Democratic party on another, does the set {Republican, Democrat} represent a sample space for the experiment? Explain.

4.2 THE CONCEPT OF PROBABILITY

The word "probability" refers to the likelihood of a given event occurring. Most people merely make subjective evaluations of situations, which amount to little more than guesses. Often, however, probabilities can be determined mathematically with considerable precision.

One method commonly used for determining probabilities is to repeat an experiment several times, recording the outcome of each experiment, and predict future outcomes as a projection of those that have already occurred. Where N represents the number of times the experiment is performed and f represents the frequency, that is, the number of times a given event actually occurs, then the probability of the event occurring on future trials is expressed as the *relative frequency, f/N*.

Example

If a pair of dice is rolled ten times and the sum of the dice is 7 on four of those rolls, then the relative frequency of the event of obtaining 7 is $f/N = 4/10$. Therefore, one may, according to these results, expect 7 to appear as a result on 4/10 of future rolls.

By performing an experiment many times, we find that the relative frequency is far from being consistent, especially for small values of N. Thus, the usefulness of relative frequency as a measure of probability is extremely limited. When N becomes very large, however, the relative frequency of an event tends to stabilize. For example, f/N may be 4/10 when N is 10, 5/20 when N is 20, 8/50 when N is 50, and 17/100 when N is 100. This suggests that there exists some number P, which the relative frequency of an event approaches as N becomes very large. This number P, when associated with an event A, is called *the probability of A* and is denoted $P(A)$.

When all sample points of a sample space are equally likely outcomes, then each sample point becomes a *simple, equally likely* event. The probability of events consisting of such sample points can be defined in the following way.

definition 4.1

If a finite sample space S consists of $n(S)$ simple, equally likely events, then the probability of A, denoted $P(A)$, is defined by the equation

$$P(A) = \frac{n(A)}{n(S)}.$$

In other words, where all of the sample points in a sample space are equally likely to occur, then the probability of any given event occurring is the number of sample points in the event, divided by the number of sample points in the sample space.

Example

The sample space of the two-coin experiment is {HH,HT,TH,TT} and consists of 4 simple, equally likely events. Therefore, if A is the event that the two coins show the same, then $A = \{HH,TT\}$, and the probability of A is

$$P(A) = \frac{n(A)}{n(S)} = \frac{2}{4} = \frac{1}{2}.$$

An experiment familiar to most people consists of tossing two ordinary six-sided dice and observing the numbers of dots on their upper faces. Table 4.1 shows the sample space that lists all possible sample points of the experiment. The first component of each ordered pair represents the first dice, and the second component represents the second dice (assuming they are distinguishable by red and green colors, respectively). Each row of the table represents a fixed value for the first dice, and each column represents a fixed value for the second dice.

From Table 4.1, the probability that the sum of the dice will be 5 is 4/36 or 1/9, because there are 4 sample points in that subset, or event, such that the sum of the dice is 5, and 36 sample points in the entire sample space. The probability of the event of at least one of the dice showing a 6 is 11/36, because 11 of the 36 sample points show 6 on at least one dice. The probability of both dice showing a 6 is 1/36. This table will be referred to in several examples and exercises throughout the chapter, so the page should be marked for future reference.

TABLE 4.1

SAMPLE SPACE FOR A TWO-DICE EXPERIMENT

Red	Green					
	1	2	3	4	5	6
1	(1,1)	(1,2)	(1,3)	(1,4)	(1,5)	(1,6)
2	(2,1)	(2,2)	(2,3)	(2,4)	(2,5)	(2,6)
3	(3,1)	(3,2)	(3,3)	(3,4)	(3,5)	(3,6)
4	(4,1)	(4,2)	(4,3)	(4,4)	(4,5)	(4,6)
5	(5,1)	(5,2)	(5,3)	(5,4)	(5,5)	(5,6)
6	(6,1)	(6,2)	(6,3)	(6,4)	(6,5)	(6,6)

In Definition 4.1, we have defined the probability of an event A in the finite sample space S to be $n(A)/n(S)$. Since the cardinality of any set must be 0 or greater, both $n(A)$ and $n(S)$ must be positive or zero, and since a sample space cannot be empty, $n(S) \neq 0$, so

$$\frac{n(A)}{n(S)} \geq 0.$$

Also, $A \subset S$ implies that $n(A) \leq n(S)$; therefore,

$$\frac{n(A)}{n(S)} \leq 1.$$

Thus, we have the following theorem, which guarantees that no probability may be less than 0 or greater than 1. $P(A) = 0$ means that A *cannot* occur, and $P(A) = 1$ means that A *must* occur.

theorem 4.1

> For any event A in a finite sample space S,

$$0 \leqslant P(A) \leqslant 1.$$

theorem 4.2

> For any finite sample space S,

$$P(S) = 1.$$

theorem 4.3

> For any event A in a finite sample space S,
> (i)   if $P(A) = 0$, then $A = \varnothing$;
> (ii) if $P(A) = 1$, then $A = S$.

Exercise Set 4.2

1.   Consider the spinner containing 8 equally likely sample points.

   Determine the probability of the event that the needle stops at
   (a) an odd integer
   (b) an integer less than 6
   (c) an integer greater than 5
   (d) an integer between 3 and 7
   (e) an integer less than 3
   (f) an integer less than 3 or greater than 6
   (g) an integer less than 10
   (h) an integer greater than 13

2.   In the experiment of tossing three fair coins, assign a probability
   to the given event:
   (a) All three coins show the same.
   (b) Not all three coins show the same.

(c) Exactly two coins show the same.

(d) No two coins show the same.

(e) At least one coin shows a head.

(f) At least one of the first two coins shows a head.

(g) At least two coins show tails.

(h) The first and the last coins show tails.

(i) The second coin shows a head.

(j) The first and third coins show the same.

3. Given an ordinary deck of 52 playing cards with 4 suits (hearts, spades, clubs, and diamonds) and 13 cards of each suit (ace, 2, 3, 4, 5, 6, 7, 8, 9, 10, jack, queen, and king),[1] and the experiment of drawing one card at random, determine the probability that the card is

(a) a heart

(b) a queen

(c) an ace

(d) a diamond

(e) a black card

(f) a red card

(g) a face card

(h) not a heart

(i) a red face card

(j) a black jack

(k) a red 3

(l) a card between 4 and 8

(m) a 10 or a face card

(n) an ace or a face card

(o) a card less than 7

(p) a card greater than 3

(q) a card greater than 4

(r) a card less than 10

(s) a black card between 7 and 10

(t) a red card between 3 and jack

4. The integers from 1 to 15 inclusive are painted on 15 balls, one number per ball. One of these balls is drawn at random. Find the probability that the number on the ball is

(a) even

(b) odd

(c) a multiple of 3

(d) a multiple of 5

(e) 4

(f) a prime number

(g) 4 or 7

(h) less than 5

(i) greater than 2

(j) greater than 7

(k) a 2-digit number

[1] Consider an ace to be 1, a jack to be 11, a queen to be 12, and a king to be 13 when questions of greater than or less than are asked.

(l) less than 8

(m) greater than 2 *and* less than 8

(n) greater than 5 *and* less than 10

(o) greater than 2 *or* less than 8

(p) greater than 5 *or* less than 10

(q) a prime number that is 2 more than another prime number

(r) less than 16

(s) an even prime number

(t) greater than 15

5. Consider the experiment of tossing two fair dice (refer to Table 4.1) and observing the number of dots on the face showing upward. What is the probability that

(a) the red dice shows a 3;

(b) the green dice shows a 5;

(c) their sum is 6;

(d) their sum is 4;

(e) their sum is 2;

(f) their sum is 5;

(g) their sum is 8;

(h) their sum is 9;

(i) their sum is 7;

(j) their sum is less than 7;

(k) their sum is greater than 7;

(l) their sum is 11;

(m) their sum is 3;

(n) their sum is 12;

(o) the red dice shows less than the green dice;

(p) the green dice shows twice as much as the red dice;

(q) one dice shows twice as much as the other;

(r) their sum is even;

(s) their sum is 7, and the red dice is less than the green dice;

(t) their sum is less than 7, and the red dice shows a 4.

4.3 INDEPENDENT EVENTS

Events may be classified into many different classifications, such as simple or compound events, events that contain equally likely outcomes or that do not, and events that are mutually exclusive or those that are not. All of these, however, may also be classified as either *in-*

dependent events or *conditional* events. If events are independent, then the outcome of one does not affect the probability of the outcome of the others. For example, in tossing an unbiased coin, the probability that the coin will come up heads is $\frac{1}{2}$ for any given toss, regardless of the outcome of previous tosses. On the other hand, consider drawing three cards from a standard deck of playing cards without replacing each card or shuffling the deck after each card is drawn, where we want to know the probability of drawing three red cards. The probability of a red card on the second draw is dependent upon the outcome of the first draw, and the probability of a red card on the third draw is dependent upon the outcome of both the first and the second draws. P(red on first draw) = 26/52, P(red on second draw) = 25/51, and P(red on third draw) = 24/50. The three probabilities are all different. Thus, we call such events *conditional* events. In this section, we restrict ourselves to the probability of independent events.

definition 4.2 (independent events)

> Any events A and B in a finite sample space S are said to be *independent* if and only if
>
> $$P(A \cap B) = P(A) \cdot P(B).$$
>
> Events that are not independent are called *dependent* events.

This definition of independent and dependent events not only gives a precise method of determining whether or not events are independent, but it also gives a method for determining the probability of the conjunction of known independent events without counting sample points. Thus, in computing probabilities, we associate the conjunction "and" with multiplication.

Example

(i) If $P(A) = 0.7$, $P(B) = 0.3$, and $P(A \cap B) = 0.25$, then

$$P(A) \cdot P(B) = (0.7)(0.3) = 0.21 \neq P(A \cap B).$$

So A and B are not independent events.

(ii) If A and B are independent events, $P(A) = \frac{2}{3}$, and $P(B) = \frac{1}{2}$, then $P(A \cap B) = P(A) \cdot P(B) = \frac{2}{3} \cdot \frac{1}{2} = \frac{1}{3}$.

Example

In the experiment of two independent spinners where A is the event "Spinner 1 stops on 2" and B is the event "Spinner 2 stops on 5":

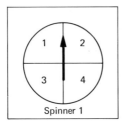
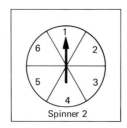

(i) The probability that Spinner 1 stops on 2 is $P(A) = \frac{1}{4}$.

(ii) The probability that Spinner 2 stops on 5 is $P(B) = \frac{1}{6}$.

(iii) The probability that Spinner 1 stops on 2 *and* Spinner 2 stops on 5 is $P(A \cap B) = P(A) \cdot P(B) = \frac{1}{4} \cdot \frac{1}{6} = \frac{1}{24}$.

Just as the probability of the conjunction of independent events can be computed in terms of multiplication of rational numbers, the probability of their disjunction, or union, can be computed in terms of addition of rational numbers. This is due to the theorem concerning the cardinality of the union of sets, which states that

$$n(A \cup B) = n(A) + n(B) - n(A \cap B).$$

A direct result of this theorem is the following theorem about probabilities. This theorem is valid for dependent as well as independent events.

theorem 4.4

For any events A and B,

$$P(A \cup B) = P(A) + P(B) - P(A \cap B).$$

Example

If A and B are independent events such that $P(A) = \frac{1}{2}$ and $P(B) = \frac{1}{3}$, then

(i) $P(A \cap B) = P(A) \cdot P(B) = \frac{1}{2} \cdot \frac{1}{3} = \frac{1}{6}$.

(ii) $P(A \cup B) = P(A) + P(B) - P(A \cap B) = \frac{1}{2} + \frac{1}{3} - \frac{1}{6} = \frac{4}{6} = \frac{2}{3}$.

Example

If $P(A) = \frac{1}{2}$, $P(B) = \frac{3}{10}$, and $P(A \cap B) = \frac{1}{6}$, then $P(A) \cdot P(B) = \frac{1}{2} \cdot \frac{3}{10} = \frac{3}{20} \neq P(A \cap B)$. So A and B are not independent events. We can still compute $P(A \cup B)$, however:

$$P(A \cup B) = P(A) + P(B) - P(A \cap B)$$
$$= \frac{1}{2} + \frac{3}{10} - \frac{1}{6} = \frac{4}{5} - \frac{1}{6} = \frac{19}{30}.$$

When A and B are mutually exclusive events, $A \cap B = \varnothing$ and $P(A \cap B) = 0$. Thus, we obtain the following corollary to Theorem 4.4.

corollary 1

If A and B are mutually exclusive events,

$$P(A \cup B) = P(A) + P(B).$$

Example

One representative is to be selected at random from a group of students consisting of 5 seniors, 4 juniors, 3 sophomores, and 7 freshmen. Let A represent the event that the representative is a senior, B the event that he is a junior, C that he is a sophomore, and D that he is a freshman. The probability that he is a senior or a freshman is

$$P(A \cup B) = P(A) + P(B) = \frac{5}{19} + \frac{7}{19} = \frac{12}{19}.$$

This may also be computed from Theorem 4.4; that is,

$$P(A \cup B) = P(A) + P(B) - P(A \cap B) = \frac{5}{19} + \frac{7}{19} - 0 = \frac{12}{19}.$$

We have shown that $A \cup \bar{A} = U$ for any set A in universe U. Therefore, for any event A in sample space S, $A \cup \bar{A} = S$. Since $P(S) = 1$ and A and $\bar{A}$ are mutually disjoint sets, Corollary 1 holds, and we have

$$P(A \cup \bar{A}) = P(A) + P(\bar{A}) = P(S) = 1.$$

The equation $P(A) + P(\bar{A}) = 1$ is equivalent to $P(A) = 1 - P(\bar{A})$, which gives us a formula which is very useful in the study of probability.

theorem 4.5

For any event A in a sample space S,

$$P(A) = 1 - P(\bar{A}),$$

and

$$P(\bar{A}) = 1 - P(A).$$

Example

In the three-coin experiment, the probability that one coin differs from the other two is

$$1 - P(\text{all three coins the same}) = 1 - \tfrac{1}{4} = \tfrac{3}{4}$$

Combining Theorem 4.5 with DeMorgan's laws, we obtain the following corollary.

corollary 2

For any events A and B in sample space S,
(i) $P(A \cup B) = 1 - P(\overline{A \cup B}) = 1 - P(\bar{A} \cap \bar{B})$;
(ii) $P(A \cap B) = 1 - P(\overline{A \cap B}) = 1 - P(\bar{A} \cap \bar{B})$.

The reason this result is so useful is that it provides a way of computing the probability of the union of events in terms of intersection, which is usually much less complicated to use. The second part of the corollary is usually much less useful, since it allows us to compute the probability of an intersection of events in terms of a union of events, which merely complicates the issue.

Example

In the two-dice experiment,
(i) the probability that at least one dice shows a 3 is
$1 - P(\text{neither dice shows a 3})$
$= 1 - P(\text{the first dice does not show a 3 } and \text{ the second dice does not show a 3})$

$$1 - \frac{5}{6} \cdot \frac{5}{6} = 1 - \frac{25}{36} = \frac{11}{36}.$$

(ii) the probability that both dice show 6 may be computed in
the usual manner; that is,

$$P(6 \text{ on first and } 6 \text{ on second}) = P(6 \text{ on first}) \cdot P(6 \text{ on second})$$

$$= \frac{1}{6} \cdot \frac{1}{6} = \frac{1}{36}.$$

This probability may also be determined using Corollary 2 in
the following way, although this tends to complicate this
type of problem.

$P(6 \text{ on first and } 6 \text{ on second})$

$$= 1 - P(\overline{6 \text{ on first and } 6 \text{ on second}})$$
$$= 1 - P(6 \text{ not on first or } 6 \text{ not on second})$$
$$= 1 - [P(6 \text{ not on first}) + P(6 \text{ not on second})$$
$$- P(6 \text{ not on first and not on second})]$$
$$= 1 - \left(\frac{5}{6} + \frac{5}{6} - \frac{5}{6} \cdot \frac{5}{6}\right)$$
$$= 1 - \left(\frac{10}{6} - \frac{25}{36}\right)$$
$$= 1 - \frac{35}{36}$$
$$= \frac{1}{36}.$$

Exercise Set 4.3

1. Let an experiment consist of making several spins of the
needle of the spinner.

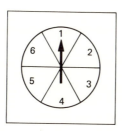

(a) What is the probability that the first spin stops on 3?
(b) What is the probability that the first two spins both end on 5

(that is, the first spin stops on 5, *and* the second spin stops on 5)?

(c) What is the probability that at least one of the first three spins ends on 4 (that is, the first spin ends on 4, *or* the second spin ends on 4, *or* the third spin ends on 4)?

(d) What is the probability that exactly one (that is, one and only one) of the first three spins ends on 4?

(e) What is the probability that the first spin ends on 1, and the second spin ends on 2?

(f) What is the probability that the first spin ends on 1, *or* the second spin ends on 2?

(g) What is the probability that the first spin ends on an even integer, *and* the second spin ends on an odd integer?

(h) What is the probability that the first spin ends on an even integer, *or* the second spin ends on an odd integer?

(i) What is the probability that the first spin ends on 1, the second on 2, the third on 3, the fourth on 4, the fifth on 5, and the sixth on 6?

(j) What is the probability that none of the first three spins ends on an even integer?

2. Given that A and B are independent events in a sample space such that $P(A) = \frac{3}{7}$ and $P(B) = \frac{2}{7}$, find the following probabilities.

(a) $P(A \cap B)$ (f) $P(\overline{A \cup B})$

(b) $P(A \cup B)$ (g) $P(\bar{A} \cap B)$

(c) $P(\bar{A})$ (h) $P(\bar{A} \cup B)$

(d) $P(\bar{B})$ (i) $P(A \cap \bar{B})$

(e) $P(\overline{A \cap B})$ (j) $P(A \cup \bar{B})$

3. A bag contains 6 black marbles, 4 red marbles, and 2 white marbles. The marbles are drawn in succession, with each marble being replaced before the next is drawn. Find each of the following:

(a) P(The first marble is black, the second red, and the third white)

(b) P(The first marble is white)

(c) P(The first two marbles drawn are black)

(d) P(The first three marbles drawn are all red)

(e) P(At least one of the first two marbles drawn is red)

(f) P(At least one of the first two marbles drawn is white)

(g) P(Exactly one of the first two marbles drawn is black)

(h) P(None of the first three marbles drawn is red)

(i) P(None of the first three marbles drawn is black)

(j) P(The first three marbles drawn are all of the same color)

4. In the two-dice experiment whose sample space is illustrated in Table 4.1, find the following probabilities. Where the disjunction "or" or the phrase "at least" is used, use the theorem $P(A \cup B) = 1 - P(\bar{A} \cap \bar{B})$ to compute the probability.

   (a) What is the probability that one dice shows a 4 and the other dice shows an even number?

   (b) What is the probability that the red dice shows less than 3 and the green dice shows more than 4?

   (c) What is the probability that the red dice shows odd and the green shows less than 4?

   (d) What is the probability that the red dice shows odd or the green shows less than 4?

   (e) What is the probability that at least one dice shows 4?

   (f) What is the probability that at least one dice shows odd?

   (g) What is the probability that exactly one dice shows 4?

   (h) What is the probability that the green dice shows the same as the red dice?

   (i) What is the probability that neither dice shows 5?

   (j) What is the probability that the dice do not show the same?

5. Must independent events be mutually exclusive? Explain and give an example.

6. Must mutually exclusive events be independent? Explain and give an example.

4.4 CONDITIONAL PROBABILITY

Now that we have considered methods of computing probabilities for independent events, we extend our discussion to include dependent events as well. Probabilities computed for this more general case are called *conditional probabilities*.

definition 4.3 (conditional probability)

For any events A and B in sample space S, the conditional probability of B, given A, is denoted by $P(B|A)$ and is defined by the equation

$$P(B \mid A) = \frac{P(A \cap B)}{P(A)}$$

if $P(A) \neq 0$.

Example

In the experiment of drawing cards in succession without replacement, from a standard deck of playing cards,

(i) the probability that the second card is red, given that the first card is black is 26/51, since 26 of the remaining 51 cards are black;

(ii) the probability that the third card is a heart given that neither of the first two cards are hearts is 13/50, since 13 of the remaining 50 cards are hearts;

(iii) the probability that the second card is a queen given that the first card is a queen is 3/51, or 1/17, since 3 of the remaining 51 cards are queens.

From Definition 4.3 we obtain the equation

$$P(A \cap B) = P(A) \cdot P(B \mid A)$$

which gives a way of computing $P(A \cap B)$ whether or not A and B are independent events. If A and B are independent events, then $P(B \mid A) = P(B)$ because the very nature of independent events dictates that knowledge of one outcome does not affect the probability of another outcome. Thus, the equation $P(A \cap B) = P(A) \cdot P(B)$ is merely a special case of the more general formula $P(A \cap B) = P(A) \cdot P(B \mid A)$.

The other formulas used in the previous section were already given in the general case, but are still affected by Definition 4.3. For example,

$$P(A \cup B) = P(A) + P(B) - P(A \cap B)$$

holds for all cases, but it should be remembered that this equation becomes

$$P(A \cup B) = P(A) + P(B) - P(A) \cdot P(B \mid A).$$

Likewise, the equations

$$P(A) = 1 - P(\bar{A})$$
$$P(A \cap B) = 1 - P(\bar{A} \cup \bar{B})$$
$$P(A \cup B) = 1 - P(\bar{A} \cap \bar{B})$$

hold for all cases, and the last equation becomes

$$P(A \cup B) = 1 - P(\bar{A}) \cdot P(\bar{B}|\bar{A}).$$

Example

Consider a bag containing 15 colored balls, which are drawn at random without replacement. Seven of the balls are green; 5 are blue; and 3 are red.

(i) The probability that the first ball drawn is red and the second is green is computed as

$$P(R \cap G) = P(R) \cdot P(G|R);$$

that is,

P(first ball red and second ball green)

$= P$(first ball red) $\cdot$ P(second ball green, given that first ball is red)

$$= \frac{3}{15} \cdot \frac{7}{14} = \frac{1}{10}.$$

(ii) The probability that at least one of the first two balls drawn is blue is computed as

$$P(A \cup B) = 1 - P(\overline{A \cup B}) = 1 - P(\bar{A} \cap \bar{B}) = 1 - P(\bar{A}) \cdot P(\bar{B}|\bar{A});$$

that is,

P(first ball blue *or* second ball blue)
  $= 1 - P$(neither of the first two balls is blue)
  $= 1 - P$(first ball not blue *and* second ball not blue)
  $= 1 - P$(first ball not blue) $\cdot$ P(second ball not blue, given first ball not blue)

$$= 1 - \frac{10}{15} \cdot \frac{9}{14} = 1 - \frac{3}{7} = \frac{4}{7}.$$

(iii) The probability that exactly one of the first two balls drawn is green is computed as $P(A \cup B) = P(A) + P(B) - P(A \cap B) = P(A) + P(B)$ because $A \cap B$ cannot happen so $P(A \cap B) = 0$; that is,

P(first ball green and second ball not green *or* first ball not green and second ball green)
  $= P$(first ball green and second ball not green) $+ P$(first ball not green and second ball green)
  $= P$(first ball green) $\cdot P$(second ball not green, given first ball green) $+ P$(first ball not green) $\cdot P$(second ball green, given first ball not green)
$$= \left(\frac{7}{15} \cdot \frac{8}{14}\right) + \left(\frac{8}{15} \cdot \frac{7}{14}\right) = \frac{4}{15} + \frac{4}{15} = \frac{8}{15}.$$

Even though the theorems for computing probabilities are stated in terms of two events only, they can be extended to any number of events by computing from left to right, using the following general theorems.

theorem 4.6

For any events $A_1, A_2, \ldots, A_k$ in sample space S,
(i) $P(A_1 \cap A_2 \cap \cdots \cap A_k)$
  $= P(A_1) \cdot P(A_2 | A_1) \cdot P(A_3 | A_1 \cap A_2) \cdots$
    $P(A_k | A_1 \cap A_2 \cap A_3 \cap \cdots \cap A_{k-1})$;
(ii) $P(A_1 \cup A_2 \cup \cdots \cup A_k)$
  $= 1 - P(\bar{A}_1 \cap \bar{A}_2 \cap \cdots \cap \bar{A}_k).$

Example

Consider the experiment of drawing three cards from a standard deck without replacement.
(i) Let A be the event that the first card is a heart, B be the event that the second card is a heart, and C be the event that the third card is a heart.

P(first card heart *and* second card heart *and* third card heart) $= P(A \cap B \cap C) = P(A) \cdot P(B | A) \cdot P(C | A \cap B)$
$$= \left(\frac{13}{52}\right) \cdot \left(\frac{12}{51}\right) \cdot \left(\frac{11}{50}\right) = \frac{11}{850}.$$

(ii) The probability that the first six cards drawn are all black is

$$\left(\frac{26}{52}\right) \cdot \left(\frac{25}{51}\right) \cdot \left(\frac{24}{50}\right) \cdot \left(\frac{23}{49}\right) \cdot \left(\frac{22}{48}\right) \cdot \left(\frac{21}{47}\right) = \frac{253}{22{,}372}.$$

(iii) The probability that at least one of the first four cards drawn is black is

$$1 - P(\text{none of the first four cards drawn is black})$$
$$= 1 - \left(\frac{26}{52}\right) \cdot \left(\frac{25}{51}\right) \cdot \left(\frac{24}{50}\right) \cdot \left(\frac{23}{49}\right) = 1 - \frac{46}{833} = \frac{787}{833}.$$

Theorem 4.6 is valid whether the events are independent or dependent, whether the individual outcomes are equally likely or not, and whether the events occur at the same time or in succession.

Exercise Set 4.4

1. A bag contains 10 marbles—5 black, 3 red, and 2 white. An experiment consists of drawing marbles out of the bag in succession without replacement.
    (a) What is the probability of drawing a white marble second, given that a red marble was drawn first?
    (b) What is the probability of drawing a white marble second, given that a white marble was drawn first?
    (c) What is the probability of drawing a black marble second, given that a black marble was drawn first?
    (d) What is the probability of drawing a black marble on *both* of the first two drawings?
    (e) What is the probability of drawing a red marble first *and* a white marble second?

2. One card is drawn from a standard deck of 52 cards. Find the probability that it is
    (a) a spade
    (b) a spade or an ace
    (c) a diamond, given that it is red
    (d) a diamond, given that it is not a club
    (e) an ace, given that it is a spade
    (f) a queen and a heart
    (g) a king and a diamond

(h) a black face card

(i) a 3, given that it is red

(j) less than 5, given that it is not a face card

(k) a heart and a face card

(l) not a black face card

3. Two cards are drawn in succession without replacement from a standard deck of 52 cards. Find the probability that

(a) both are black kings;

(b) both are face cards;

(c) the second card is a spade, given that the first card is a spade;

(d) both cards are spades;

(e) both cards are red;

(f) both are hearts;

(g) the first is a heart, and the second is a spade;

(h) one is a heart, and the other is a spade (that is, the first is a heart and the second is a spade, or the first is a spade and the second is a heart);

(i) the second is a king, given that the first is a queen;

(j) the second is a king, given that the first is a king;

(k) the second card is a 10 of diamonds, given that the first card is a diamond other than 10;

(l) the first card is a king, and the second is a queen;

(m) the first card is black, and the second card is red;

(n) both cards are of the same suit;

(o) both cards have the same numeral.

4. Answer the questions in Problem 3, assuming that each card is replaced before the next card is drawn.

5. In the two-dice experiment (whose sample space is illustrated by Table 4.1), what is the probability that

(a) the sum of the dice is 7, given that the red dice shows 4;

(b) the sum of the dice is 11, given that the red dice does not show 6;

(c) the green dice shows less than the red dice, given that they do not show the same;

(d) the red dice shows 3, and the green dice shows less than 3;

(e) the two dice show the same;

(f) the red dice shows less than 3, and the green dice shows more than 3;

(g) the red dice shows a 4, and the sum of the dice is less than 8;

(h) the red dice shows one half as much as the green dice;

(i) the green dice shows less than 4, given that the sum of the dice is 6;

(j) the sum of the dice is one more than the red dice.

6. A bag contains 3 black marbles, 4 red marbles, 5 white marbles, and 6 green marbles. They are drawn in succession without replacement. What is the probability that

(a) the second is black, given that the first is red;

(b) the third is green, given that the first two are green;

(c) the third is red, given that the first two are white;

(d) the third is red, given that the first two are black;

(e) the first three are black;

(f) the first is green, the second white, and the third red;

(g) the first two are black, and the third is green;

(h) the first five are all white;

(i) at least one of the first two is green;

(j) neither of the first two is green;

(k) exactly one of the first two is black;

(l) exactly one of the first three is red;

(m) the first is green, the second black, the third red, and the fourth white;

(n) at least one of the first three is white;

(o) at least one of the first four is red;

(p) at least one of the first thirteen is green.

7. A box contains 50 red balls, 25 green balls, and 25 yellow balls, thoroughly mixed. Three balls are drawn from the box with replacement after each draw. What is the probability that

(a) all three balls are green;

(b) the first ball is red, the second is green, and the third is yellow;

(c) at least one of the balls is green;

(d) the third ball is yellow, given that the first two are yellow;

(e) none of the balls is red;

(f) the first and third balls are red, and the second is green;

(g) exactly one of the three balls is red;

(h) the third ball is green, given that the first two are red;

(i) the three balls are all of the same color;

(j) the three balls are all of different colors.

8. Answer each of the questions in Problem 7 if the balls are drawn without replacement.

4.5 BAYES' THEOREM

Where H_1, H_2, H_3, . . . , H_n are events that partition a sample space S, we often wish to determine the effect of some arbitrary event E on the probability of one or more of these events individually. For example, we may wish to determine $P(H_1 | E)$. To determine a method of computing such a probability, let us consider some of the information that we have. First, we know that the elements of a partition are mutually exclusive and that their union is the universe, or sample space, from which they come; that is,

$$S = H_1 \cup H_2 \cup H_3 \cup \cdots \cup H_n$$

so

$$S \cap E = (H_1 \cup H_2 \cup H_3 \cup \cdots \cup H_n) \cap E.$$

Since

$$E \subset S, S \cap E = E = (H_1 \cap E) \cup (H_2 \cap E) \cup \cdots \cup (H_n \cap E).$$

Therefore,

$$P(E) = P[(H_1 \cap E) \cup (H_2 \cap E) \\ \cup (H_3 \cap E) \cup \cdots \cup (H_n \cap E)].$$

Since H_1, H_2, . . . , H_n are mutually disjoint, so also are $(H_1 \cap E)$, $(H_2 \cap E)$, . . . , $(H_n \cap E)$. Thus, by Theorem 4.4, Corollary 1, $P(E) = P(H_1 \cap E) + P(H_2 \cap E) + \cdots + P(H_n \cap E)$. We can now use this result in the conditional probability equation

$$P(H_1 | E) = \frac{P(H_1 \cap E)}{P(E)}$$

to obtain the following theorem, called *Bayes' theorem*.

theorem 4.7 (Bayes' theorem)

For any events $H_1, H_2, H_3, \ldots, H_n$ that partition sample space S and an arbitrary event E in S such that $P(E) \neq 0$,

$$P(H_1 \mid E) = \frac{P(H_1 \cap E)}{P(H_1 \cap E) + P(H_2 \cap E) + \cdots + P(H_n \cap E)}$$

and similar results hold for H_2, H_3, H_4, and so on, to H_n.

Example

Three women are sewing coats in a factory, and woman A sews 25 percent of the coats produced each week, woman B sews 40 percent, and woman C sews 35 percent. In a week's work, 1 percent of the coats produced by woman A, 3 percent of those produced by woman B, and 2 percent of those produced by woman C are defective. A coat is drawn at random from a weekly output and is found to be defective. What is the probability that the defective coat was (a) sewn by woman A? (b) sewn by woman B? (c) sewn by woman C?

solution:

Let H_1 represent the event that the coat was sewn by woman A,

H_2 represent the event that the coat was sewn by woman B,

H_3 represent the event that the coat was sewn by woman C, and

E represent the event that the coat is defective.

Before we can apply Bayes' theorem to determine $P(H_1 \mid E)$, $P(H_2 \mid E)$, and $P(H_3 \mid E)$, we must determine $P(H_1 \cap E)$, $P(H_2 \cap E)$, and $P(H_3 \cap E)$.

$$P(H_1 \cap E) = P(\text{woman A sewed the coat } and \text{ the coat is defective})$$
$$= P(H_1) \cdot P(E \mid H_1) = (25 \text{ percent})(1 \text{ percent})$$
$$= 0.0025.$$
$$P(H_2 \cap E) = P(\text{woman B sewed the coat } and \text{ the coat is defective})$$
$$= P(H_2) \cdot P(E \mid H_2) = (40 \text{ percent})(3 \text{ percent})$$
$$= 0.012.$$

$P(H_3 \cap E) = P$(woman C sewed the coat *and* the coat is defective)

$= P(H_3) \cdot P(E|H_3) = $ (35 percent)(2 percent)

$= 0.007.$

We can now use the results to compute

(a) $P(H_1|E) = \dfrac{P(H_1 \cap E)}{P(H_1 \cap E) + P(H_2 \cap E) + P(H_3 \cap E)}$

$= \dfrac{0.0025}{(0.0025) + (0.012) + (0.007)}$

$= \dfrac{.0025}{0.0215} = \dfrac{25}{215} = \dfrac{5}{43}.$

(b) $P(H_2|E) = \dfrac{P(H_2 \cap E)}{P(H_1 \cap E) + P(H_2 \cap E) + P(H_3 \cap E)}$

$= \dfrac{0.012}{0.0215} = \dfrac{120}{215} = \dfrac{24}{43}.$

(c) $P(H_3|E) = \dfrac{P(H_2 \cap E)}{P(H_1 \cap E) + P(H_2 \cap E) + P(H_3 \cap E)}$

$= \dfrac{0.007}{0.0215} = \dfrac{70}{215} = \dfrac{14}{43}.$

In the previous example, note that both

$$P(H_1) + P(H_2) + P(H_3) = 0.25 + 0.40 + 0.35 = 1$$

and

$$P(H_1|E) + P(H_2|E) + P(H_3|E) = \frac{5}{43} + \frac{24}{43} + \frac{14}{43} = 1.$$

The probabilities $P(H_1)$, $P(H_2)$, and $P(H_3)$ are called *a priori* probabilities *before* the information that the item is defective. The probabilities $P(H_1|E)$, $P(H_2|E)$, and $P(H_3|E)$ are called *a posteriori* probabilities, that is, the probabilities *after* the information that the item is defective.

Exercise Set 4.5

1. One representative is chosen at random from a group of 10 freshmen (2 girls and 8 boys), 17 sophomores (12 girls and 5

boys), 20 juniors (10 girls and 10 boys), and 23 seniors (12 girls and 11 boys). What is the probability that

(a) the representative is a senior;
(b) the representative is a sophomore;
(c) the representative is a female;
(d) the representative is a male senior;
(e) the representative is a female freshman;
(f) the representative is freshman or female;
(g) the representative is junior or male;
(h) the representative is a senior, given that he is male;
(i) the representative is male, given that he is a senior;
(j) the representative is a junior, given that she is female;
(k) the representative is a junior, given that he is male;
(l) the representative is a freshman, given that she is female;
(m) the representative is an upperclassman, given that he is male;
(n) the representative is not a female freshman;
(o) the representative is not a female or a freshman.

2. Four women wash the dishes in the school cafeteria. Woman A washes 30 percent of the dishes, while woman B washes 10 percent, woman C washes 40 percent, and woman D washes 20 percent. Woman A breaks 1 percent of the dishes she washes each day, while woman B breaks 2 percent, woman C breaks 3 percent, and woman D breaks 5 percent. If a broken dish is found, what is the probability that it was broken by

(a) woman A?
(b) woman B?
(c) woman C?
(d) woman D?

3. In a lightbulb factory, machine A produces 15 percent of the bulbs produced per day, machine B produces 22 percent, machine C produces 18 percent, machine D produces 30 percent, and machine E produces 15 percent. The rate of their defective bulbs is 2 percent, 5 percent, 5 percent, 2 percent, and 4 percent, respectively. A bulb is drawn from a day's production at random.

(a) What is the probability that the bulb is defective?
(b) What is the probability that the bulb is defective and produced by machine A?
(c) What is the probability that the bulb is defective and produced by machine B?

(d) What is the probability that the bulb is produced by machine C and is not defective?

(e) What is the probability that the bulb is either produced by machine D or is defective?

(f) If the bulb is defective, what is the probability that it was produced by machine A?

(g) If the bulb is defective, what is the probability that it was produced by machine B?

(h) If the bulb is defective, what is the probability that it was produced by machine C?

(i) If the bulb is defective, what is the probability that it was produced by machine D?

(j) If the bulb is defective, what is the probability that it was produced by machine E?

(k) If the bulb is *not* defective, what is the probability that it was produced by machine D?

(l) If the bulb is *not* defective, what is the probability that it was *not* produced by machine C?

4. In a certain class at Dulsville High School, the girls are classified into three categories—shy, friendly, and forward. Since there are the same number of girls as boys in the school, all of the girls have a date for the junior prom, and all are asked at sometime between 2 weeks and 2 hours prior to the dance. The following table gives the probability of each type of girl being asked at various times over this time interval. In this class, 25 percent of the girls are forward, 25 percent friendly, and 50 percent shy. The name of a girl from this class is drawn at random.

Type of Girl	Time Interval During Which Asked (Interval does not include upper limit.)		
	5 days to 2 weeks	1 day to 5 days	2 hours to 1 day
Forward	0.80	0.19	0.01
Friendly	0.20	0.60	0.20
Shy	0.00	0.21	0.79

(a) If she was asked to the dance 8 days before the dance, what is the probability that she is friendly?

(b) What is the probability that she is friendly and is asked out at least 5 days before the dance?

(c) What is the probability that she is shy and is asked out at least 5 days before the dance?
(d) Given that she is asked out 2 hours before the dance, what is the probability that she is shy?
(e) Given that she is asked out 3 days before the dance, what is the probability that she is forward?

4.6 BERNOULLI TRIALS

Often we deal with an experiment that is repeated over and over under the same conditions, such as drawing cards, marbles, and so on, with replacement. If such a sequence of independent repeated trials has exactly two possible outcomes to each trial, which we label S for success and F for failure, then the trials are called *Bernoulli trials* in honor of James Bernoulli, who investigated such situations around the year 1700.

To determine the probability of exactly K successes in n Bernoulli trials, let $P(S) = p$ be the probability of success on each trial and $P(F) = 1 - p = q$ be the probability of failure on each trial. Since the experiments are independent, the probability $P(X)$ of exactly k successes and $n - k$ failures in any given sequence is the product

$$P(X) = p^k q^{n-k}.$$

The number of such sequences possible is the number of combinations of n things taken k at a time. Thus, we obtain the following theorem.

theorem 4.8

The probability of exactly k successes in a sequence of n Bernoulli trials with $P(S) = p$ and $P(F) = q$ is denoted by $b(k,n,p)$ and is given by the equation

$$b(k,n,p) = C(n,k)p^k q^{n-k} = \binom{n}{k} p^k q^{n-k}.$$

Example

An urn contains 15 white balls and 5 black balls, which are drawn one at a time, with each being replaced and all of them

mixed before the next is drawn. What is the probability that exactly 3 of the first 5 balls drawn are white?

solution:
P(exactly 3 of the first 5 balls are white) $= b(3,5,\frac{3}{4})$

$$b\left(3,5,\frac{3}{4}\right) = \binom{5}{3}\left(\frac{3}{4}\right)^3\left(\frac{1}{4}\right)^2 = \left(\frac{5!}{2!3!}\right)\left(\frac{27}{64}\right)\left(\frac{1}{16}\right) = \frac{135}{512}.$$

Example

An experiment consists of drawing one card at a time from a
standard deck of cards, with each card being replaced and the
deck shuffled before the next card is drawn. (a) What is the probability that *exactly* 2 of the first 4 cards drawn are hearts? (b)
What is the probability that *at least* 2 of the first 4 cards are
hearts?

solution:

(a) P(exactly 2 hearts in 4 cards) $= b\left(2,4,\frac{1}{4}\right)$

$$= \binom{4}{2}\left(\frac{1}{4}\right)^2\left(\frac{3}{4}\right)^2 = \frac{27}{128}.$$

(b) P(at least 2 hearts in 4 cards)
   $= P$(exactly 2 hearts in 4 cards or exactly 3 hearts in 4 cards
   or exactly 4 hearts in 4 cards)

$$= b\left(2,4,\frac{1}{4}\right) + b\left(3,4,\frac{1}{4}\right) + b\left(4,4,\frac{1}{4}\right)$$

$$= \binom{4}{2}\left(\frac{1}{4}\right)^2\left(\frac{3}{4}\right)^2 + \binom{4}{3}\left(\frac{1}{4}\right)^3\left(\frac{3}{4}\right)^1 + \binom{4}{4}\left(\frac{1}{4}\right)^4\left(\frac{3}{4}\right)^0$$

$$= \frac{27}{128} + \frac{3}{64} + \frac{1}{256} = \frac{67}{256}.$$

Exercise Set 4.6

1. The probability of your guessing the correct answer is $\frac{3}{5}$ for each
   question of a 20-question true-false test. Set up the following
   problems.
   (a) What is the probability that you will answer exactly 10 questions correctly?

(b) What is the probability that you will answer exactly 14 questions correctly?

(c) What is the probability that you will answer at least 15 questions correctly?

(d) What is the probability that you will answer at least 18 questions correctly?

(e) What is the probability that you will answer no questions correctly?

2. A door-to-door salesman has 1 chance in 5 of making a sale at each door he knocks on. (Just set the problem up.)

(a) What is the probability that he will make exactly 10 sales after knocking on 50 doors?

(b) What is the probability that he makes exactly 10 sales in knocking on 100 doors?

(c) What is the probability that he makes at least 1 sale in knocking on 5 doors?

(d) What is the probability that he makes no sales in knocking on 50 doors?

3. If the probability of winning a hand in a particular card game is $\frac{2}{5}$, what is the probability of

(a) winning 3 out of 3 hands?

(b) winning exactly 2 out of 5 hands?

(c) winning at least 2 out of 5 hands?

(d) losing 5 out of 5 hands?

(e) winning exactly 3 out of 5 hands?

4. Find the number of hands that must be played in the card game referred to in Problem 3 to be 95 percent certain of winning at least one hand.

5. In six tosses of an unbiased coin, what is the probability of obtaining exactly 3 heads and 3 tails?

6. If the probability of any one person being involved in an automobile accident this year is $\frac{1}{4}$ for each citizen of this country:

(a) What is the probability that exactly 1 member in a family of 6 will be involved in such an accident?

(b) What is the probability that at least 1 member in a family of 6 will be involved in such an accident?

(c) What is the probability that at least 1 person in a town of 60 will be involved in such an accident?

4.7 SUMMARY

In this chapter, we have discussed methods of computing probabilities of many different types of events. We found that one method of computing probabilities of an event consisting of equally likely sample points is merely to count the sample points in the event and the sample points in the sample space and express the probability of the event as their ratio; that is,

$$P(A) = \frac{n(A)}{n(S)}.$$

We found, however, that this method of computing probabilities is both extremely limited in value as well as cumbersome to work with. Thus, we developed the following theorems, which are valid whether or not the events being considered are independent, whether or not they are mutually exclusive, and whether or not their sample points are equally likely.

1) $P(A \cap B) = P(A) \cdot P(B \mid A)$
   (This becomes $P(A \cap B) = P(A) \cdot P(B)$ if A and B are independent events.)
2) $P(A \cup B) = P(A) + P(B) - P(A \cap B)$
   (This becomes $P(A \cup B) = P(A) + P(B)$ if A and B are mutually exclusive.)
3) $P(A) = 1 - P(\bar{A})$
4) $P(A \cup B) = 1 - P(\overline{A \cup B}) = 1 - P(\bar{A} \cap \bar{B})$
5) If $H_1, H_2, H_3, \ldots, H_n$ are events that determine a partition of the sample space S, then we can determine $P(H_i \mid E)$ for any arbitrary event E of S, using the following equation (Bayes' theorem)

$$P(H_i \mid E) = \frac{P(H_i \cap E)}{P(H_1 \cap E) + P(H_2 \cap E) + \cdots + P(H_n \cap E)}.$$

6) Finally, when a sequence of independent repeated trials is performed under identical conditions such that each trial has only two possible outcomes, S and F, the probability

$b(k,n,p)$ of exactly k successes in n trials can be computed according to the equation

$$b(k,n,p) = \binom{n}{k} p^k q^{n-k}$$

where $q = 1 - p$.

Review Exercise Set 4.7

1. Consider the experiment of rolling one regular six-sided dice and tossing an unbiased coin.
    (a) What is the probability that the dice is even and the coin a head?
    (b) What is the probability that the dice is even or the coin is a head?
    (c) What is the probability that the dice shows prime?
    (d) What is the probability that the coin shows a head and the dice shows less than 5, or that the coin shows a tail and the dice shows greater than 4?
    (e) What is the probability that both the dice and the coin show the same for at least two of the first three tosses?
2. Consider the experiment of drawing a card from a standard deck of cards and replacing each card and shuffling the deck before the next card is drawn. Find the probability that
    (a) the first 3 cards are all red;
    (b) none of the first 3 cards drawn is red;
    (c) at least 1 of the first 3 cards drawn is red;
    (d) exactly 2 of the first 3 cards drawn are red;
    (e) at least 2 of the first 3 cards drawn are red;
    (f) the first card is a heart, given that it is not a spade;
    (g) the first 3 cards are all of the same color;
    (h) the first card is a heart and the second a spade;
    (i) the first card is a heart or the second a spade;
    (j) the first 3 cards are all of a different suit;
    (k) exactly 3 of the first 7 cards are black;
    (l) no 2 of the first 3 cards are of the same color;
    (m) at least 3 of the first 4 cards show the same numeral;
    (n) the first card is a 2, the second a 4, and the third an 8;
    (o) four of the first 5 cards drawn are aces.

3. Find the probabilities indicated in Problem 2 when the cards are drawn without replacement.

4. A college faculty consists of 25 percent Republicans, 35 percent Democrats, and 40 percent Independents. Two percent of the Republicans, 7 percent of the Democrats, and 15 percent of the Independents failed to vote in the last presidential election. If one faculty member is chosen at random,

    (a) find the probability that he did not vote;
    (b) find the probability that he is a Democrat and did vote;
    (c) find the probability that he is a Republican and did not vote;
    (d) find the probability that he is an Independent who did not vote;
    (e) find the probability that he is a Democrat *or* did not vote;
    (f) given that he did not vote, find the probability that he is a Republican;
    (g) given that he did not vote, find the probability that he is a Democrat;
    (h) given that he did not vote, find the probability that he is an Independent;
    (i) given that he did vote, find the probability that he is not an Independent.

chapter

5

AN INTRODUCTION
TO STATISTICS

5.1 INTRODUCTION

In Chapter 4, we discussed probability of events involving cards, dice, flipping coins, random drawings, spinners, and other experiments, the nature of which we understand thoroughly. It does not take a great deal of scientific knowledge to predict that an unbiased coin will show heads as a result of approximately $\frac{1}{2}$ of the number of times it is tossed. It is somewhat more difficult, however, to predict the outcome of human behavior in any given situation. This is due to the fact that we never know all of the factors affecting human behavior, so we cannot determine the probability of such behavior precisely as we have done with the objective experiments with which we have been working. Thus, when making predictions concerning people or other factors of which we lack a complete understanding, we collect data of similar known experiments and make predictions based on their outcomes. In this way, predictions involving a large number of cases are frequently made by studying data obtained from a smaller number of cases, called a sample. The collection and organization of such data as well as the predictions made from them fall into the field of study called *statistics*.

5.2 ORGANIZING STATISTICAL DATA

Virtually every person has a need at some time to collect and organize statistical data. The manager of a supermarket needs to know shoppers' habits in order to know what items to have on his shelves and how many checkers he will need on hand at various hours of the day each day of the week. An invester would like to know stock market trends before investing in the market. A politician needs to know voter reaction to both personal traits and political stands in order to improve his chances of being elected. A college student can be assisted greatly in selecting a major field of study if he knows employment trends. The list could go on to include every walk of life to some extent. The greater the extent is, the greater is the need to organize and tabulate the data in a way that can easily be interpreted.

Much statistical data can effectively be organized into a *table,* a *graph,* or a *frequency distribution.* For example, Table 5.1, showing the sample space for a two-dice experiment, becomes much more useful than a random tabulation of the same sample space, due to the way in which the data are organized. Note that all outcomes with a particular numeral showing on the first dice lie in the same row, those with a particular numeral showing on the second dice lie in the same column, and those whose sum is a particular number all lie on the same diagonal line. Thus, one can readily acquire much information from the very way in which the table is organized without having to actually count sample points. For instance, we can see that the sum of the dice is more likely to be 7 and less likely to be 2 or 12 than any other particular number. Likewise, we can see that the probability of the sum being 6 is the same as that of being 8, the probability of the sum being 5 is the same as that of being 9, and so on.

TABLE 5.1

	1	2	3	4	5	6
1	(1,1)	(1,2)	(1,3)	(1,4)	(1,5)	(1,6)
2	(2,1)	(2,2)	(2,3)	(2,4)	(2,5)	(2,6)
3	(3,1)	(3,2)	(3,3)	(3,4)	(3,5)	(3,6)
4	(4,1)	(4,2)	(4,3)	(4,4)	(4,5)	(4,6)
5	(5,1)	(5,2)	(5,3)	(5,4)	(5,5)	(5,6)
6	(6,1)	(6,2)	(6,3)	(6,4)	(6,5)	(6,6)

Another useful method of organizing statistical data is to classify them according to numerical measure. Such an organization of data is appropriately called a *frequency distribution.*

Example

The following are scores on a mathematics examination: 60, 70, 90, 64, 48, 74, 44, 84, 76, 36, 96, 74, 68, 92, 80, 88, 84, 80, 72, 74, 44, 64, 78, 88, 92, 32, 80, 84, 82, 80. We can express the general results of the examination with the following frequency distribution:

Test Interval	Frequency	Test Interval	Frequency
96–100	1	61–65	2
91–95	2	56–60	1
86–90	3	51–55	0
81–85	4	46–50	1
76–80	6	41–45	2
71–75	4	36–40	1
66–70	2	31–35	1

The same data may be expressed in graphical form with a *bar graph,* or *histogram,* as shown in Figure 5.1.

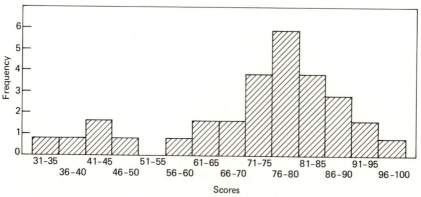

FIGURE 5.1

A third method of expressing the same statistical data is by means of a *frequency polygon*, which is obtained by joining the mid-points of the tops of the rectangles of the histogram by straight-line segments. (See Figure 5.2.)

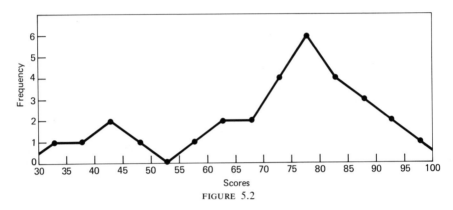

FIGURE 5.2

Exercise Set 5.1

1. Given the test scores
   53, 84, 47, 98, 61, 96, 75, 89, 77, 65, 83, 79, 69, 67, 76, 72, 85, 77, 55, 65, 88, 80, 91, 88, 77, 89, 74, 69, 69, 44, 44, 41, 83, 95, 93, 75, 93, 83, 95, 84, 65, 42, 91, 89, 85, 42, 99, 87, 77, 93, 100, 82, 64, 79, 95, 62, 49, 70, 71, 99, 42.
   (a) Make a frequency distribution with intervals of 5 points each.
   (b) Construct a histogram illustrating this information.
   (c) Construct a frequency polygon illustrating this information.

2. The following are the amounts (rounded to the nearest dollar) paid for textbooks by a class of college students:
   35, 33, 26, 52, 23, 36, 38, 30, 27, 43, 41, 50, 29, 26, 31, 34, 42, 37, 24, 36, 48, 43, 39, 49, 41, 43, 46, 29, 38, 41, 50.
   (a) Present the data as a frequency distribution with intervals of $5 each, beginning at $20.
   (b) Present the data as a histogram.
   (c) Present the data as a frequency polynomial.

3. The following are final grades given in a course:
   D, A, C+, B−, C, C, A, B+, B, C+, C, A, B, C+, C, C−, C, A−, C+, F, A−, C+, B+, B, C+, C, C−, C, B+, C+, C, C−, F, C, B+, B, C+, C, C−, B−, C, D, B, B−, C, C−, D, C, B, C, C−, D+, C−, D, D−, C−, D+, F.

(a) Present the data as a frequency distribution according to grade.
(b) Present the data as a histogram.
(c) Present the data as a frequency polynomial.

5.3 MEASURES OF CENTRAL TENDENCY

A frequency distribution table is an orderly means of arranging data, and graphs aid in visualizing and understanding the data, but both have very limited value when used alone. To further understanding of data, we often, either in conjunction with tables and graphs or alone, use *measures of central tendencies.* The most common of these measures are the arithmetic mean, the median, and the mode.

The *arithmetic mean,* which is often called the arithmetic average, is the sum of the measures, or scores, divided by the number of measures.

Example

(i) If the weights in pounds of six girls in a given apartment are 120, 109, 141, 103, 132, and 115, then their *average weight,* or *mean weight,* is

$$\frac{(120 + 109 + 141 + 103 + 132 + 115)}{6} = \frac{720}{6} = 120.$$

(ii) The following frequency distribution represents the grades obtained by a class on a certain test.

Score	Frequency
100	2
90	5
80	7
70	15
60	14
50	3
40	1

The arithmetic mean of the class on the test is computed as $m = S/N$, where m is the arithmetic mean; S is the sum of the numbers obtained by finding the product of each score and the frequency with which it occurred; and N is the number of scores.

Score	Frequency	Product
100	2	200
90	5	450
80	7	560
70	15	1050
60	14	840
50	3	150
40	1	40

so $m = S/N = 3290/47 = 70$.

The term *median,* when referring to a measure of central tendency, refers to the score of the "middle" student; that is, half of the subjects obtain scores that are less than or equal to that score. If the total frequency N is even, then there is no "middle" score, so we take the score of the lowest subject in the upper half of the distribution to be the median score.[1] In the second part of the previous example, there were 47 students who took an examination. The median score M is the score of the 24th student, that is, the student with a score lower than or equal to the scores of 23 students and higher than or equal to the scores of 23 students. Thus, $M = 70$. The fact that $m = 70$ and $M = 70$ is purely coincidental, because often the mean and median are not the same.

Example

(i) Given the scores

57, 63, 92, 71, 86, 47, 73, 81, 76, 20, 60,

we find the median by first listing the scores in decreasing order.

$$
\begin{array}{l}
92 \\
86 \\
81 \\
76 \\
73 \\
71 \quad \leftarrow \quad \text{median} \\
63 \\
60 \\
57 \\
47 \\
20
\end{array}
$$

[1] Some take the average of the two middle scores to be the median.

Since there are 5 scores below 71, and 5 scores above 71, $M = 71$.

(ii) Given the frequency distribution

Score	Frequency
15	3
12	9
11	7
10	4
9	3
7	1
5	2
2	3

we find that $N = 32$. Thus, the median score M is that of the 16th entry from the top, which is one of the 7 entries obtaining a score of 11; that is, $M = 11$.

The third measure of central tendency is called the *mode,* which is that score or interval of scores obtained by the greatest number of entries, if such a score exists. In Part (ii) of the previous example, the mode is 12, since 9 entries obtained a score of 12, and no other score was obtained by that many entries.

The concepts of mean, median, and mode are all useful. While the mean is the most commonly used measure of central tendency, it can be misleading when there is an extreme measure or score at one end of the scale with nothing to balance at the other end. In such cases, the median is often a more practical measure to use. On the other hand, the mode shows the most popular measure or score, thus telling the most about the greatest number of entries. All three of these measures of central tendencies are often referred to as "averages." Thus, when an average is asked for, there is often a need to ask "which average?"

Example

If the incomes of the families on Elm Street are

$2000, $3000, $4000, $5000, and $36,000, then $m = \$10,000$, and $M = \$4000$.

If one were to use the mean as the average, then he would conclude that the average income on that street is $10,000, yet only one of the five families on the street has that large an income. Thus, the median of $4000 gives a much more accurate indication of the income level of the people on that street.

Example

Jerry went to a store to purchase a thermometer. Upon examining those available, he noted that of the six that were for sale, one registered 87°; three registered 75°; one registered 130°; and one registered 80°. The mean reading is 87°; the median is 80°; and the mode is 75°. Which average is the best indication of the correct temperature? In this case, the mode is the best indication of accuracy.

Exercise Set 5.2

1. Given the following as points made by each player in a basketball game
   14, 26, 36, 9, 8, 23, 8, 17, 2, 2, 8, 1.
   (a) Find the arithmetic mean.
   (b) Find the median score.
   (c) Find the mode.

2. The following are outside temperatures in degrees, taken at noon on each day of two consecutive weeks:
   78, 80, 79, 85, 78, 87, 86, 89, 91, 85, 78, 74, 60, 54.
   (a) Find the arithmetic mean.
   (b) Find the median.
   (c) Find the mode.

3. The age at the time of marriage of each student of a particular graduating class is shown by the following frequency distribution.

Age	Frequency
18	5
19	9
20	12
21	15
22	17
23	10
24	8
25	5
26	6
27	4
30	3
45	1
62	2

   (a) Find the arithmetic mean.
   (b) Find the median.
   (c) Find the mode.

4. Given the following frequency distribution of IQ scores

Score	Frequency
126–130	8
121–125	15
116–120	25
111–115	30
106–110	20
101–105	25
96–100	10
91–95	7

   (a) Find the arithmetic mean. (For 126–130 use 128, for 121–125 use 123, and so on.)
   (b) Find the median.
   (c) Find the mode.

5. The following are scores on an examination:
   80, 90, 55, 75, 60, 85, 70, 55, 80, 70, 70, 55, 50, 70, 80, 70, 20, 80, 95, 75, 95, 85, 95, 85, 60, 85, 80, 70, 55, 60, 75, 80, 70, 85, 80, 90, 100, 80, 95, 85, 80, 80, 70, 60, 80, 60, 55, 90, 85, 80, 100, 55, 80, 90, 100, 85, 75, 75, 95, 75, 85, 90, 90, 55, 85, 65.

   (a) Arrange them into a frequency distribution with intervals of 5, that is, 96–100, 91–95, and so on.
   (b) Find the arithmetic mean.
   (c) Find the median.
   (d) Find the mode.

6. An automobile manufacturer advertises that his cars average 25 miles per gallon of gasoline. Would you be more impressed if the average were the mean, the median, or the mode? Explain.

7. In estimating the amount of income tax to be collected from a given community, which average taxable income – the mean, the median, or the mode – would be most helpful to know? Explain.

5.4 VARIABILITY

In the previous section, we discussed measures of central tendencies, or averages, but all such measures are extremely limited in the information they give. For example, consider two college basketball teams in the same conference, both of whose starting teams average 75 points per game, with the 75 points representing the mean, the median, and the mode scores. From this information, it appears that the teams are comparable on offense. Yet suppose the individual player averages of the 5 starters on one team are 15 points, 16 points, 16 points, 14 points, and 14 points, respectively, while the individual averages of the starters on the other team are 34 points, 5 points, 2 points, 9 points, and 25 points, respectively. Certainly, the teams could no longer be considered comparable, and furthermore, this added information becomes extremely valuable to a team preparing to play either of these teams. This new information has to do with the concept of *dispersion* or *variability,* which is one of the most important and useful concepts studied in statistics.

There are several measures of variability that are used. One of the most common such measures is the *range,* which is the difference in the largest score and the smallest score. For example, the ranges of player averages on the basketball teams just mentioned are 2 points for the first team and 32 points for the second team. This gives some indication of team balance, but not as accurate an indication as one would like. For example, there could be a third team in the conference with player averages of 2 points, 13 points, 13 points, 13 points, and 34 points, respectively, thus having a team average of 75 points and a range of 32 points, yet having far better scoring balance than the second team mentioned.

A second measure of variability is called the *average deviation,* denoted AD. The average deviation is merely the arithmetic mean of the absolute values of the differences of the scores from the mean, that is, the positive deviation from the mean.

Example

(i)   The average deviation of the scores 2, 5, 9, 25, and 34 can be found as follows (*Note:* $x = X - m$):

Score (X)	Deviation from the Mean ($|x|$)
2	13
5	10
9	6
25	10
34	19
(Sum of the deviations, denoted $\Sigma|x|$)	= 58

$$AD = \frac{\Sigma|x|}{N} \qquad \text{(the sum of the deviations divided by the number of scores)}$$

$$= \frac{58}{5} = 11.6.$$

(ii) The average deviation of the scores 2, 13, 13, 13, and 34 can be computed as follows:

Score (X)	Deviation from the Mean ($|x|$)
2	13
13	2
13	2
13	2
34	19
	38

$$AD = \frac{\Sigma|x|}{N} = \frac{38}{5} = 7.6.$$

Although the range and the average deviation are relatively easy to compute, the most stable index of variability is an index that we call the *standard deviation*, commonly denoted by either SD or the Greek letter sigma σ. The standard deviation is the square root of the variance V, which can be computed by the following three steps:

1) Find the deviation (or difference) of each score from the mean.
2) Square each deviation.
3) Find the mean of the squared deviations.

Symbolically, we write the formulas

$$V = \frac{\Sigma x^2}{N}$$

$$\sigma = SD = \sqrt{V}.$$

Example

(i) To find the variance and standard deviation of the scores 2, 5, 9, 25, and 34, we compute from the table as follows:

Score (X)		Deviation (x)	x^2
2		-13	169
5	$m = 15$	-10	100
9	$N = 5$	-6	36
25		10	100
34		19	361
			$\Sigma x^2 = 766$

$$V = \frac{\Sigma x^2}{N} = \frac{766}{5} = 153.2$$

$$\sigma = \text{SD} = \sqrt{V} = \sqrt{153.2} \approx 12.37.$$

(ii) To find the variance and standard deviation of the scores 2, 13, 13, 13, and 34, we compute from the table as follows:

Score (X)	Deviation (x)	x^2	
2	−13	169	
13	−2	4	$m = 15$
13	−2	4	$N = 5$
13	−2	4	
34	19	361	
		$\Sigma x^2 = 542$	

$$V = \frac{\Sigma x^2}{N} = \frac{542}{5} = 108.4$$

$$\sigma = \text{SD} = \sqrt{V} = \sqrt{108.4} \approx 10.39.$$

(*Note:* To compute the square root of a number, see the tables in the appendix.)

(iii) To find the variance and standard deviation of the scores 14, 14, 15, 16, and 16, we compute from the table as follows:

Score (X)	Deviation (x)	x^2	
14	−1	1	
14	−1	1	$m = 15$
15	0	0	$N = 5$
16	1	1	
16	1	1	
		$\Sigma x^2 = 4$	

$$V = \frac{\Sigma x^2}{N} = \frac{4}{5} = 0.8$$

$$\sigma = \text{SD} = \sqrt{V} = \sqrt{0.8} \approx 0.89.$$

(iv) To find the variance and standard deviation of the scores 16, 7, 4, 8, 5, 21, 20, and 23, we compute from the table as follows:

Score (X)	Deviation (x)	x^2	
16	3	9	
7	−6	36	
4	−9	81	$m = \dfrac{104}{8} = 13$
8	−5	25	
5	−8	64	$N = 8$
21	8	64	
20	7	49	
23	10	100	
104		$\Sigma x^2 = 428$	

$$V = \frac{\Sigma x^2}{N} = \frac{428}{8} = 53.5$$

$$\sigma = \text{SD} = \sqrt{V} = \sqrt{53.5} \approx 7.28.$$

Exercise Set 5.3

1. The following are temperatures recorded for a given week in Centerville: 65°, 73°, 70°, 68°, 63°, 64°, and 45°.
   (a) Find the mean.
   (b) Find the range.
   (c) Find the average deviation.
   (d) Find the variance.
   (e) Find the standard deviation.

2. The married members of the Goodrich family have annual incomes of $6000, $4700, $3000, $6500, $8700, and $14,000, respectively.
   (a) Find the mean.
   (b) Find the median.
   (c) Find the range.
   (d) Find the average deviation.
   (e) Find the standard deviation.

3. The 5 boys in the Bunker family have the following heights: 6'3", 6'1", 6'9", 5'11", and 6'.
   (a) Find their mean height.
   (b) Find their median height.
   (c) Find the range in their height.
   (d) Find the average deviation.
   (e) Find the standard deviation.

4. On a particular English test, the men in the class have a mean score of 70.6 percent and a standard deviation of 21, and the women have a mean score of 70.2 percent and a standard deviation of 5. From this information, how would you compare the performance of the men in the class with that of the women?

5.5 THE NORMAL DISTRIBUTION

In the tabulation and interpretation of statistical data, the mean, median, and mode are often found to be the same. When this is the case, it is possible that a graph of the data is the *normal probability curve* or simply the *normal curve*. This particular curve represents the most common and most important distribution of data used in statistical analysis. Data that determine a normal distribution have a mean, median, and mode that are the same, and a graph such as that in Figure 5.3, which is a symmetrical "bell-shaped" curve with 68 percent of the area under the curve lying within 1 SD of the mean, 95 percent of that area within 2 SD of the mean, and 99.7 percent of that area within 3 SD of the mean.

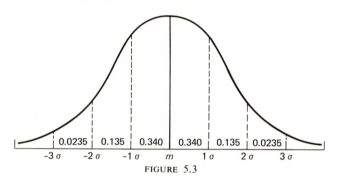

FIGURE 5.3

The condition necessary for a curve to be a normal curve allows many such curves to have the same mean with different standard deviations, as in Figure 5.4, or the same standard deviation with different means. Thus, the normal curve is not unique.

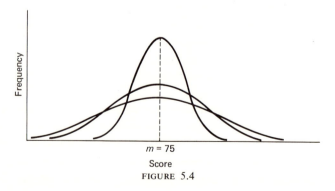

FIGURE 5.4

The normal curve becomes very important to probability theory, because probability of events can be computed in terms of area under this curve. To do so, however, we first convert raw scores from the data to *standard scores,* which we also call z scores. Where X is a raw score, m is the mean score, and σ is the standard deviation, we convert the raw score to the standard z score by the formula

$$z = \frac{X - m}{\sigma}$$

In other words, a standard score is merely the directed distance in standard deviations that a score lies from the mean.

Once a standard score is determined, the area under the curve up to that point can be found in Table C.2 of Appendix C. The area under the entire curve is considered to be 1, since this represents the probability of the entire sample space. Thus, the area under the curve between any two points represents the probability that a randomly selected score will be between those two points.

Example

The grades obtained on a mathematics examination have a normal distribution with a mean of 68 and a standard deviation of 10. If one test paper is randomly drawn from the results, what is the probability that the score on the paper is above 60?

solution:

A score of 60 is a standard score of $z = (60 - 68)/10 = -0.8$. According to Table C.2 in Appendix C, the probability that the score is below this point is 0.2119. Therefore, the probability that the score is above this point is $1 - 0.2119 = 0.7881$.

Example

The bushels per acre of wheat produced by wheat farmers in a particular county provide a normal distribution, with a mean of 30 bushels per acre. If a particular acre of wheat is chosen at random, what is the probability that the yield is between 20 bushels per acre and 35 bushels per acre?

solution:

A yield of 35 bushels per acre has a standard score of

$$z_1 = \frac{35 - 30}{6} = \frac{5}{6} \approx 0.83$$

and a yield of 20 bushels per acre has a standard score of

$$z_2 = \frac{20 - 30}{6} = \frac{-10}{6} \approx -1.67.$$

Therefore, the probability of the yield being between z_1 and z_2 is

$$P(z_1) - P(z_2) = 0.7967 - 0.0475 = 0.7492.$$

Example

The average IQ score for the workers in a given industry was found to be 105 with a standard deviation of 10. There are 200 workers in the business, and their IQ scores are normally distributed.

(a) How many workers would be expected to have IQ scores between 100 and 120?

(b) How many workers would be expected to have IQ scores below 80?

(c) How many workers would be expected to have IQ scores above 120?

solutions:

(a) An IQ score of 120 has a standard score of

$$z_1 = \frac{120 - 105}{10} = 1.5.$$

An IQ score of 100 has a standard score of

$$z_2 = \frac{100 - 105}{10} = -0.5.$$

The probability that an individual chosen at random has an IQ between 100 and 120 is

$$P(z_1) - P(z_2) = 0.9332 - 0.3085 = 0.6247.$$

Therefore, 62.47 percent of the workers have IQ's in this range; that is,

$$(62.47 \text{ percent})(200) = 124.94 \approx 125$$

workers with IQ's between 100 and 120.

(b) An IQ score of 80 has a standard score of $z = (80 - 105)/10 = -2.5$. Using Table C.2, we find the probability of an IQ being below a standard score of -2.5 is 0.0062. Therefore, 0.62 percent of the workers have IQ's in this range.

$$(0.62 \text{ percent})(200) = 1.24 \approx 1$$

worker with an IQ below 80.

(c) An IQ score of 120 has a standard score of $z = (120 - 105)/10 = 1.5$. The probability of a score being above a z score is

$$1 - P(z) = 1 - 0.9332 = 0.0668.$$

Therefore, 6.68 percent of the workers have IQ's in this range.

$$(6.68)(200) = 13.36 \approx 13$$

workers with IQ's above 120.

Exercise Set 5.4

1. Find the area under the normal curve below the standard score of
   (a) 3.0
   (b) 2.7
   (c) 0.4
   (d) 1.3
   (e) 3.2
   (f) -1.3

(g) −2.1   (l) −3.13
(h) −0.5   (m) 1.79
(i) −0.9   (n) 2.38
(j) −0.02   (o) −2.65
(k) 0.52

2. Find the area under the normal curve between the standard scores of
   (a) 2.3 and 0.1   (f) −2.31 and 0.13
   (b) 0.53 and 3.15   (g) 3.42 and −1.23
   (c) 1.05 and 1.50   (h) 3.04 and −2.14
   (d) 3.01 and 0.13   (i) −2.49 and −3.17
   (e) 2.47 and 1.35   (j) −2.63 and −1.58

3. In a normal distribution with a mean score of 42 and a standard deviation of 12, find the standard score that corresponds to each of the following raw scores.
   (a) 53   (i)   100
   (b) 72   (j)   37
   (c) 90   (k)   20
   (d) 31   (l)   5
   (e) 26   (m) 42
   (f) 40   (n) 70
   (g) 87   (o) 39
   (h) 93

4. In a normal distribution with a mean score of 60 and a standard deviation of 10, find the probability that a score selected at random is
   (a) below 40   (f) between 50 and 70
   (b) below 50   (g) between 40 and 90
   (c) above 70   (h) between 30 and 50
   (d) above 90   (i) between 80 and 90
   (e) above 80   (j) between 50 and 65

5. The scores on an English examination are normally distributed with a mean of 78 and a standard deviation of 7. One examination paper is chosen at random.
   (a) What is the probability that the score is below 50?
   (b) What is the probability that the score is above 70?
   (c) What is the probability that the score is between 60 and 80?

6. The average yearly family income in Cokerville is $8000, with a

standard deviation of $2000. Assume that the town has 100 families, and their incomes are normally distributed. Answer the following questions to the nearest whole number.

(a) How many families have incomes in excess of $10,000 per year?

(b) How many families have incomes between $7000 and $10,000 per year?

(c) How many families have incomes of less than $3000 per year?

7. Assume that the life span of power saws manufactured by a given company is normally distributed, with a mean of 5 years and standard deviation of 1.5 years. If a 1-year unconditional replacement warranty is issued with each saw sold, and 12,000 are sold in 1 year, how many of those saws will need to be replaced under the warranty?

8. The manufacturer in Problem 7 wishes to issue as long a warranty as is economically feasible with his product. He finds that he can afford to replace 4 percent of the saws sold. How long a period of time then (to the nearest month) should the warranty cover?

5.6 CORRELATION

In previous sections of this chapter, we have discussed methods of organizing and analyzing data concerning performance in some defined trait or variable. Often, however, it is more important to know the relationship of two or more variables to each other than to know about each one independently. For example, we may wish to know if success in marriage is related to age at the time of marriage, if success in business is related to IQ scores, if criminal tendencies are related to poverty, or if political party affiliation is related to education. We call such relationships *correlations,* and the extent of the relation can be measured by a *correlation coefficient.*

When the relationship between two variables is "linear," that is, its graph is a straight line, the correlation can be expressed by the *product-moment* correlation coefficient. This coefficient may be thought of primarily as the ratio that expresses the extent to which changes in one variable are related to or dependent upon changes in another variable. A formula for computing this correlation coefficient, denoted by r, is

$$r = \frac{\Sigma(xy)}{N\sigma_x\sigma_y}$$

where x is the deviation from the mean of the first variable; y is the deviation from the mean of the second variable; σ_x is the standard deviation of the first variable; σ_y is the standard deviation of the second variable; and N is the number of subjects in the sample. This is by no means the only correlation coefficient used, but it is the most common correlation coefficient used and the one that we restrict ourselves to in this book.

Example

Five college seniors chosen at random have the following IQ scores and grade-point averages (GPA):

Student	IQ (X)	GPA (Y)
A	120	3.25
B	112	2.05
C	143	3.90
D	110	3.45
E	115	2.85

The mean IQ score is 120, and the mean GPA is 3.10. Thus, we are able to construct the following table of information needed to compute the standard deviation of IQ scores σ_x, the standard deviation of grade-point averages σ_y, and the product-moment correlation r. The symbol x represents the deviation from the mean IQ, and the symbol y represents the deviation from the mean GPA.

Student	IQ (X)	GPA (Y)	x	y	xy	x^2	y^2
A	120	3.25	0	0.15	0.00	0	0.225
B	112	2.05	−8	−1.05	8.40	64	1.103
C	143	3.90	23	0.80	18.40	529	0.640
D	110	3.45	−10	0.35	−3.50	100	0.123
E	115	2.85	−5	−0.25	1.25	25	0.625
					24.55	718	2.716

$$m_x = 120, \qquad \sigma_x = \sqrt{\frac{\Sigma x^2}{N}} = \sqrt{\frac{718}{5}} \approx 12$$

$$m_y = 3.10, \qquad \sigma_y = \sqrt{\frac{y^2}{N}} = \sqrt{\frac{2.716}{5}} \approx 0.74$$

$$r = \frac{\Sigma xy}{N\sigma_x\sigma_y} = \frac{24.55}{5(12)(0.74)} = \frac{24.55}{44.4} = 0.55.$$

Example

The following table gives paired scores for six individuals. The first score is age in years, and the second score is a physical coordination score based on a scale from 1 to 10 with 10 being the best possible score.

Person	Age (X)	Coor. (Y)	x	y	xy	x^2	y^2
A	25	8	−13	2	−26	169	4
B	31	7	−7	1	−7	49	1
C	18	7	−20	1	−20	400	1
D	42	6	4	0	0	16	0
E	52	5	14	−1	−14	196	1
F	60	3	22	−3	−66	484	9
					−133	1314	16

$$m_x = 38 \text{ yr}, \qquad \sigma_x = \sqrt{\frac{\Sigma x^2}{N}} = \sqrt{\frac{1314}{6}} = \sqrt{219} \approx 14.80$$

$$m_y = 6 \text{ yr}, \qquad \sigma_y = \sqrt{\frac{\Sigma y^2}{N}} = \sqrt{\frac{16}{6}} = \sqrt{2.67} \approx 1.63$$

$$r = \frac{\Sigma xy}{N\sigma_x\sigma_y} = \frac{-133}{6(14.80)(1.63)} = \frac{-133}{144.744} \approx -0.92.$$

Correlation coefficients range over a scale from −1.00 through 0.00 to 1.00. A correlation of 1.00 indicates perfect correlation; that is, when one variable increases or decreases, the other variable increases or decreases correspondingly. They have a direct variation, and a graph of the correlation lies completely on the line $y = x$ as in Figure 5.5. A correlation of −1.00 indicates that as one variable increases, the other decreases, as in Figure 5.6. A correlation of 0.00 indicates that the variables have no consistent relationship at all. Few correlation coefficients are ever 1.00, −1.00, or 0.00, however. We consider coefficients "high" if they are near 1.00 or −1.00, and "low" if they are near 0.00.

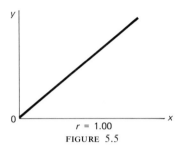

r = 1.00

FIGURE 5.5

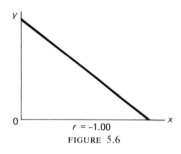

r = -1.00

FIGURE 5.6

Exercise Set 5.5

1. Seven men living on a particular street have yearly incomes and years of education beyond high school as given in the following table.

Person	Yearly Income (X)	Years of Ed. Beyond H.S. (Y)
A	$7000	0
B	$17,000	5
C	$5000	1
D	$10,000	4
E	$26,000	4
F	$20,000	6

(a) Find the mean yearly income. (Use the nearest $1000.)
(b) Find the mean education level.
(c) Find the standard deviation of the yearly income, σ_x.
(d) Find the standard deviation of the education level, σ_y.
(e) Find the product-moment correlation.

2. The following table gives paired heights and weights of five college freshmen.

Student	Height in Inches (X)	Weight in Pounds (Y)
A	68	185
B	72	170
C	70	180
D	69	165
E	66	150

(a) Find the mean height in inches.
(b) Find the mean weight in pounds.
(c) Find the deviation of each height from the mean.
(d) Find the deviation of each weight from the mean.

(e) Find the standard deviation of the heights.

(f) Find the standard deviation of the weights.

(g) Find the product-moment correlation.

3. The following table illustrates the relationship of dates per week to college grade-point average.

Student	Dates per Week	Grade-Point Average
A	1	3.9
B	0	3.5
C	5	1.7
D	2	3.5
E	3	2.8
F	1	3.7
G	6	2.1
H	2	3.2
I	4	2.3
J	7	1.0

(a) Find the mean number of dates per week. (Use the nearest whole number.)

(b) Find the mean grade-point average. (Use the nearest one tenth.)

(c) Find the standard deviation of the dates per week.

(d) Find the standard deviation of the grade-point averages.

(e) Find the product-moment correlation.

5.7 SUMMARY

In this chapter, we have discussed methods of organizing and interpreting statistical data. The most common methods of organizing such data are with tables, graphs, and frequency distributions.

Important to the interpretation of statistical data are *measures of central tendency,* the most common of which include the mean, the median, and the mode. All three are important, and although the mean is the most common measure of central tendency, there are situations in which either the median or the mode gives more accurate interpretation of data.

A second type of measure used for interpretation of data is *variability.* Measures of variability discussed are the range, the average deviation, the variance, and the standard deviation. Formulas for these are the following:

$$AD = \frac{\Sigma|x|}{N}$$

$$V = \frac{\Sigma x^2}{N}$$

$$\sigma = \sqrt{V} = \sqrt{\frac{\Sigma x^2}{N}}.$$

The mean and standard deviation of a distribution become important to many measures used to interpret data. One of these measures is a *standard score*, which can be computed when the distribution is a *normal distribution*. Such standard scores can be used to compute probabilities of events occurring with randomly selected samples. The equation used for determining a standard score is

$$z = \frac{X - m}{\sigma}.$$

When we wish to determine the extent to which one variable accompanies or is related to another variable, such a relationship can be measured by a correlation coefficient. The most common correlation coefficient is the product-moment correlation coefficient, which can be computed by the equation

$$r = \frac{\Sigma xy}{N\sigma_x\sigma_y}.$$

The following symbols are used throughout the chapter with the indicated meanings.

N: number of scores

m: mean

M: median

$\begin{cases} X \\ Y \end{cases}$: raw scores

$\begin{cases} x \\ y \end{cases}$: deviation from the mean

AD: average deviation

$\begin{cases} SD \\ \sigma \end{cases}$: standard deviation

V: variance

z: standard score

σ_x: standard deviation of variable x

σ_y: standard deviation of variable y

r: product-moment correlation coefficient

Σ: summation

Exercise Set 5.6

The following are scores obtained by students on standardized mathematics, English, and physics examinations.

Student	Math.	Eng.	Phys.
A	90	80	85
B	40	70	50
C	65	80	60
D	70	50	75
E	85	60	95
F	75	85	70
G	95	90	90
H	50	75	45
I	80	65	85
J	60	40	70

1. Find the mean of the mathematics examination.
2. Find the median of the mathematics examination.
3. Find the mode of the English examination, using a 10-point range, for each interval.
4. Find the mean of the English examination.
5. Find the median of the physics examination.
6. Find the mean of the physics examination.
7. Find the standard deviation of the mathematics examination.
8. Find the standard deviation of the English examination.
9. Find the standard deviation of the physics examination.
10. Find the average deviation of the English examination.
11. Find the range of the physics examination.
12. Find the product-moment correlation of the mathematics and English examinations.
13. Find the product-moment correlation of the physics and English examinations.

chapter

6
VECTORS
AND MATRICES

6.1 INTRODUCTION

Vectors and matrices have had numerous applications in sciences for a great number of years, but as our society has become more and more computerized, the usefulness of these concepts has increased astronomically. Where the study of vectors was once confined to the fields of physics, mathematics, and statistics, we now find as many applications to the social and biological sciences. Thus, interest in the algebra of vectors and matrices has intensified considerably over the past few years.

Many people are somewhat familiar with the concept of a vector, yet have no understanding whatever about matrices even though a vector is a matrix. Since this is the case, we choose to first consider the special type of matrices called vectors before discussing matrices in general.

This chapter is somewhat mechanical in nature rather than applied or theoretical. This is due to the fact that the mechanics of vectors and matrices are a necessary prerequisite to Chapter 7, which applies these principles; yet a complete discussion of these topics is beyond the level of this book.

6.2 VECTORS

Often a vector is thought of as a force with a given magnitude and direction. Although such a force can be studied as a vector, vectors are not limited to such a narrow definition. Many people are often surprised to find that a great number of concepts, which they had previously thought to be unrelated to vectors, are in fact vectors. In fact, any concept that can be expressed by an ordered pair, an ordered triplet, or an ordered n tuple (for any natural number n) of real numbers can be studied as a vector.

definition 6.1

For any real numbers $a_1, a_2, a_3, \ldots, a_n$ and any natural number n, the ordered n tuple $[a_1, a_2, a_3, \ldots, a_n]$ is called a *vector*, of *order n,* or an *n-dimensional vector.*

A two-dimensional vector $[a,b]$ can be illustrated graphically, as shown in Figure 6.1, either as a point, a units from the vertical axis and b units from the horizontal axis, in the Cartesian coordinate plane, or as the directed line segment from the origin to that point (or from any point to the point a units to the right of it and b units above it). The length of the segment is called the *magnitude* of the vector. If $\alpha = [a_1, a_2, a_3, \ldots, a_n]$, then the magnitude of α, denoted $|\alpha|$, is given by $|\alpha| = \sqrt{(a_1)^2 + (a_2)^2 + (a_3)^2 + \cdots + (a_n)^2}$.

Example

(i) If $\alpha = [5,-3]$ then $|\alpha| = \sqrt{5^2 + (-3)^2} = \sqrt{25+9} = \sqrt{34}$.

(ii) If $\beta = [2,-1,3]$, then $|\beta| = \sqrt{2^2 + (-1)^2 + 3^2} = \sqrt{4+1+9} = \sqrt{14}$.

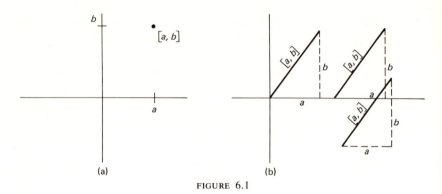

FIGURE 6.1

In a like manner, a three-dimensional vector can be illustrated graphically. A similar graph for a vector of order greater than three, however, would be quite difficult to construct. This does not mean that such vectors do not exist, or even that they are uncommon. The difficulty lies with the graph—not the vector.

We can illustrate addition of two- or three-dimensional vectors graphically by considering a vector to be a translation of a point. For example, the vector [a,b] translates each point a units to the right and b units up. Thus, the sum [a,b] + [c,d] is interpreted as first translating a point a units to the right and b units up, and then translating that point c units to the right and d units up. To illustrate this, we place the beginning of the second vector at the end of the first vector. Their sum or *resultant* is the vector from the origin to the end of the final vector, as in Figure 6.2.

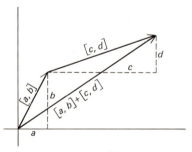

FIGURE 6.2

Example

(i) $[4,2] + [3,3] = [4 + 3, 2 + 3] = [7,5]$.

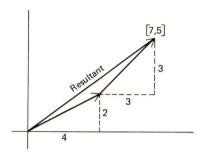

(ii) $[5,2] + [-3,4] = [5 + (-3), 2 + 4] = [2,6]$.

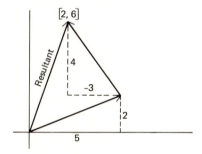

Vectors, like numbers, can be added and multiplied. The following definitions are valid for n-dimensional vectors for any natural number n, but the graphs are restricted to two dimensions, for obvious reasons.

definition 6.2 (equality of vectors)

For any vectors

$$\boldsymbol{\alpha} = [a_1, a_2, a_3, \ldots, a_n]$$
$$\boldsymbol{\beta} = [b_1, b_2, b_3, \ldots, b_n]$$

of order n, $\boldsymbol{\alpha} = \boldsymbol{\beta}$ if and only if $a_1 = b_1$, $a_2 = b_2$, $a_3 = b_3$, ..., and $a_n = b_n$; that is, vectors are equal if and only if their corresponding components are equal.

Example

    (i) $[5,3,1] \neq [5,1,3]$, because $3 \neq 1$ and $1 \neq 3$.

    (ii) $[4 + 2, \ 3] = [1 + 5, \ 3]$, because $4 + 2 = 1 + 5$ and $3 = 3$.

    (iii) $[a + b, \ c + d] = [b + a, \ d + c]$, because $a + b = b + a$ and $c + d = d + c$.

definition 6.3 (addition and subtraction of vectors)

    For any vectors $\boldsymbol{\alpha}$ and $\boldsymbol{\beta}$ of order n with

$$\boldsymbol{\alpha} = [a_1, a_2, a_3, \ . \ . \ . \ , a_n]$$
$$\boldsymbol{\beta} = [b_1, b_2, b_3, \ . \ . \ . \ , b_n]$$

    their sum is the n-dimensional vector

$$\boldsymbol{\alpha} + \boldsymbol{\beta} = [a_1 + b_1, \ a_2 + b_2, \ . \ . \ . \ , \ a_n + b_n]$$

    and their difference is the n-dimensional vector

$$\boldsymbol{\alpha} - \boldsymbol{\beta} = [a_1 - b_1, \ a_2 - b_2, \ . \ . \ . \ , \ a_n - b_n].$$

Example

    (i) $[5,3] + [7,4] = [5 + 7, \ 3 + 4] = [12,7]$.

    (ii) $[4,-3,1] - [5,1,-3] = [4 - 5, -3 - 1, 1 - (-3)] = [-1,-4,4]$.

    (iii) $[3,2,-1,4] + [-6,1,5,0] = [3 + (-6), 2 + 1, -1 + 5, 4 + 0] = [-3,3,4,4]$.

    (iv) $[5,-1,3] + [4,2]$ is undefined, because the vectors are not of the same dimension.

    A second operation, which we define on vectors, is called *scalar multiplication* — multiplication of a vector by a real number — which we call a scalar to distinguish it from a vector.

definition 6.4 (scalar multiplication)

    For any vector $\boldsymbol{\alpha} = [a_1, a_2, a_3, \ . \ . \ . \ , a_n]$ of order n and real number c, we define their product to be the n-dimensional vector

$$c\boldsymbol{\alpha} = [ca_1, ca_2, ca_3, \ . \ . \ . \ , ca_n].$$

Example
(i)   If $\alpha = [5,3,2]$, then $5\alpha = 5[5,3,2] = [25,15,10]$.
(ii) If $\beta = [-3,1]$, then $-3\beta = -3[-3,1] = [9,-3]$.
(iii) If $\mathbf{u} = [5,1,3]$ and $\mathbf{v} = [-2,4,1]$, then $2\mathbf{u} + 3\mathbf{v} = [10,2,6] + [-6,12,3] = [4,14,9]$.

Intuitively, we may think of the product of a scalar c and a vector α as the vector whose direction is the same as that of α and whose magnitude is the product of c and the magnitude of α.

definition 6.5 (dot product)
For any two vectors

$$\alpha = [a_1,a_2,a_3, \ldots ,a_n]$$
$$\beta = [b_1,b_2,b_3, \ldots ,b_n]$$

of order n, their *dot product* or *inner product* is a scalar defined by

$$\alpha \cdot \beta = a_1b_1 + a_2b_2 + a_3b_3 + \cdots + a_nb_n.$$

Example
(i)   If $\alpha = [5,1,4]$ and $\beta = [-3,2,6]$, then $\alpha \cdot \beta = 5(-3) + 1(2) + 4(6) = -15 + 2 + 24 = 11$.
(ii) If $\mathbf{u} = [9,-2]$ and $\mathbf{v} = [4,-5]$, then $\mathbf{u} \cdot \mathbf{v} = 9(4) + (-2)(-5) = 36 + 10 = 46$.

Exercise Set 6.1
1. Find the magnitude of each of the following vectors.
    (a) $[5,1]$ (i) $[4,-1,-2]$
    (b) $[4,-3]$ (j) $[0,-5,0]$
    (c) $[-2,3]$ (k) $[5,-1,2,3]$
    (d) $[6,-1]$ (l) $[-4,0,1,2]$
    (e) $[4,0]$ (m) $[0,5,0,-5]$
    (f) $[4,1,2]$ (n) $[-3,1,4,0]$
    (g) $[-1,0,3]$ (o) $[5,7,9,-3]$
    (h) $[-3,1,2]$
2. Perform the indicated operation where possible. If not possible, tell why.

$$\mathbf{u} = [5,1,3] \quad \mathbf{v} = [-2,0,6] \quad \mathbf{w} = [1,-4,2]$$

(a) $\mathbf{u} + \mathbf{v}$ (i) $(\mathbf{u} + \mathbf{v}) + \mathbf{w}$

(b) $\mathbf{u} \cdot \mathbf{v}$ (j) $\mathbf{u} \cdot \mathbf{v} \cdot \mathbf{w}$

(c) $5\mathbf{u}$ (k) $5(\mathbf{u} + \mathbf{w})$

(d) $\mathbf{u} - \mathbf{v}$ (l) $-3(\mathbf{u} \cdot \mathbf{v})$

(e) $2\mathbf{u} + \mathbf{v}$ (m) $4(\mathbf{v} \cdot \mathbf{u})$

(f) $2\mathbf{u} + 3\mathbf{w}$ (n) $2(3\mathbf{u})$

(g) $3\mathbf{u} - 4\mathbf{v}$ (o) $0(\mathbf{w})$

(h) $2\mathbf{u} \cdot \mathbf{v}$

3. Identify each of the following as true or false, where $\mathbf{u}$, $\mathbf{v}$, and $\mathbf{w}$ represent vectors of the same order. Verify those that are true, and for those that are false, either give counterexamples or explain why they are false. a, b, and c are scalars.

(a) $\mathbf{u} + \mathbf{v} = \mathbf{v} + \mathbf{u}$ (f) $\mathbf{u} \cdot (\mathbf{v} + \mathbf{w}) = \mathbf{u} \cdot \mathbf{v} + \mathbf{u} \cdot \mathbf{w}$

(b) $\mathbf{u} \cdot \mathbf{v} = \mathbf{v} \cdot \mathbf{u}$ (g) $c(\mathbf{u} + \mathbf{v}) = c\mathbf{u} + c\mathbf{v}$

(c) $(\mathbf{u} + \mathbf{v}) + \mathbf{w} = \mathbf{u} + (\mathbf{v} + \mathbf{w})$ (h) $(ab)\mathbf{u} = a(b\mathbf{u})$

(d) $(\mathbf{u} \cdot \mathbf{v}) \cdot \mathbf{w} = \mathbf{u} \cdot (\mathbf{v} \cdot \mathbf{w})$ (i) $b(\mathbf{v} \cdot \mathbf{w}) = (b\mathbf{v}) \cdot (b\mathbf{w})$

(e) $c(\mathbf{u} \cdot \mathbf{v}) = (c\mathbf{u}) \cdot \mathbf{v}$ (j) $a\mathbf{u} = \mathbf{u}a$

4. Use vector addition and scalar multiplication to verify that the diagonals of a parallelogram bisect each other. You may use Figure 6.3.

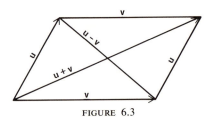

FIGURE 6.3

6.3 MATRICES

Around 1850 the English mathematician James Joseph Sylvester formed squares from rows and columns, which he extracted from a rectangular arrangement of terms and called it a *matrix*. This seemed to be an appropriate word, since its meaning is a place in which something develops or originates. Within a short time, other mathematicians began to recognize that such arrays are convenient devices for extending the concept of a number. Now entire courses are devoted

to the study of matrices, and they are used in applied problems in virtually every academic field of study.

The matrices with which we concern ourselves here are rectangular arrays of real numbers, which we call *entries*, or *elements*, of
the matrices. We display such a matrix in square brackets. The order
or dimension of a matrix is *mn*, where *m* is the number of rows, and *n*
is the number of columns that the matrix contains. We say that such a
matrix is an "*m* by *n*" matrix, denoted *m* × *n*. Thus,

$$\begin{bmatrix} 2 & 1 & 4 \\ -1 & 3 & 0 \end{bmatrix}$$

is a 2 × 3 matrix,

$$\begin{bmatrix} -4/5 & 1/2 \\ 2 & 2/3 \\ 0 & -5 \end{bmatrix}$$

is a 3 × 2 matrix, and

$$\begin{bmatrix} 5 & 10 & 0 & 1/4 \\ 12 & 0 & -14 & 31 \\ -1 & 2/3 & 1/2 & \pi \end{bmatrix}$$

is a 3 × 4 matrix.

An *m* × *m* matrix with the same number of rows as columns is
called a *square* matrix.

A matrix consisting of a single row or a single column is a vector.
Up to this point, we have denoted each vector as a single row, or a *row
vector*. We may just as well have denoted every vector as a single
column, or a *column vector*. From this point on, we shall distinguish
between row and column vectors in order to have vector operations be
consistent with matrix operations in general. The vector notation
[*a*,*b*] and [*a*,*b*,*d*] is consistent with our matrix notation for a row
vector, while a column vector is denoted

$$\begin{bmatrix} a \\ b \end{bmatrix} \quad \text{or} \quad \begin{bmatrix} a \\ b \\ c \end{bmatrix}$$

for a 2×1 and a 3×1 matrix, respectively. Since a vector is a matrix, we shall no longer treat vectors as a separate topic, but as a special case of the general topic of matrices.

We use capital letters to denote matrices and small letters to denote their entries and real numbers in general. When referring to a matrix in general, a useful convention is to denote each element as some a_{ij}, where the subscript i indicates the row of the entry, and the subscript j indicates its column. When this notation is used, we denote a general element of a matrix $\mathbf{A}$ by a_{ij} and $\mathbf{A}$ itself by $[a_{ij}]$.

$$\mathbf{A} = [a_{ij}] = \begin{bmatrix} a_{11} & a_{12} & a_{13} & a_{14} & \cdots & a_{1n} \\ a_{21} & a_{22} & a_{23} & a_{24} & \cdots & a_{2n} \\ a_{31} & a_{32} & a_{33} & a_{34} & \cdots & a_{3n} \\ a_{41} & a_{42} & a_{43} & a_{44} & \cdots & a_{4n} \\ \cdot & \cdot & \cdot & \cdot & \cdot & \cdot \\ \cdot & \cdot & \cdot & \cdot & \cdot & \cdot \\ \cdot & \cdot & \cdot & \cdot & \cdot & \cdot \\ a_{m1} & a_{m2} & a_{m3} & a_{m4} & \cdots & a_{mn} \end{bmatrix}.$$

The symbol a_{32} indicates the entry in the third row and second column, while a_{23} indicates the entry in the second row and third column. If A is square, then the diagonal consisting of the elements $a_{11}, a_{22}, a_{33}, \ldots, a_{mm}$ is called the *principal diagonal*.

Since a matrix is not itself a real number, we need definitions for both relations and operations on matrices. The first relation that we define is equality. Note the similarity of Definitions 6.6 and 6.8 to the corresponding definitions for vectors.

definition 6.6 (equality of matrices)

Two matrices $\mathbf{A}$ and $\mathbf{B}$ are said to be equal if and only if they are of the same order $m \times n$, and for every $0 < i \leqslant m$ and $0 < j \leqslant n$, $a_{ij} = b_{ij}$; that is, all corresponding entries of the two matrices are equal.

definition 6.7 (the transpose of a matrix)

The *transpose* of an $m \times n$ matrix $\mathbf{A}$ is the $n \times m$ matrix $\mathbf{A}^t$, whose rows are the corresponding columns of $\mathbf{A}$; that is, $a_{ij} = a^t_{ji}$.

Example

If

$$\mathbf{A} = \begin{bmatrix} -2 & -3 \\ 4 & 14 \\ 0 & -5 \end{bmatrix}$$

then

$$\mathbf{A}^t = \begin{bmatrix} -2 & 4 & 0 \\ -3 & 14 & -5 \end{bmatrix}.$$

definition 6.8 (addition of matrices)

The sum of two $m \times n$ matrices $\mathbf{A}$ and $\mathbf{B}$ is the $m \times n$ matrix $\mathbf{A} + \mathbf{B}$, whose entry in the ith row and the jth column is the sum of the corresponding entries in $\mathbf{A}$ and $\mathbf{B}$ for any $0 < i \leq m$ and $0 < j \leq n$; that is, $(a + b)_{ij} = a_{ij} + b_{ij}$.

Example

(i)   If

$$\mathbf{A} = \begin{bmatrix} 4 & 11 & -3 \\ 12 & -3 & 6 \end{bmatrix} \quad \text{and} \quad \mathbf{B} = \begin{bmatrix} -5 & -12 & 4 \\ -20 & 10 & -3 \end{bmatrix}$$

then

$$\mathbf{A} + \mathbf{B} = \begin{bmatrix} 4 + (-5) & 11 + (-12) & -3 + 4 \\ 12 + (-20) & -3 + 10 & 6 + (-3) \end{bmatrix}$$

$$= \begin{bmatrix} -1 & -1 & 1 \\ -8 & -7 & 3 \end{bmatrix}.$$

(ii)   If

$$\mathbf{A} = \begin{bmatrix} 2 & -1 \\ 3 & 4 \end{bmatrix} \quad \text{and} \quad \mathbf{B} = \begin{bmatrix} -2 & 1 \\ -3 & -4 \end{bmatrix}$$

then

$$\mathbf{A} + \mathbf{B} = \begin{bmatrix} 2 + (-2) & (-1) + 1 \\ 3 + (-3) & 4 + (-4) \end{bmatrix} = \begin{bmatrix} 0 & 0 \\ 0 & 0 \end{bmatrix}.$$

Even though matrices are not real numbers, matrix operations possess many, but not all, of the properties of corresponding real-number operations. We state these properties as theorems without proof, but they can be proved immediately from matrix definitions and corresponding properties of real numbers.

theorem 6.1

For any $m \times n$ matrices **A, B, C,** with real-number entries,

(i) **A + B** is an $m \times n$ matrix with real-number entries (closure)

(ii) **A + B = B + A**   (commutativity)

(iii) **(A + B) + C = A + (B + C)**   (associativity)

(iv) If **A = B,** then **A + C = B + C.**

definition 6.9 (zero matrix)

A matrix with each entry equal to 0 is called a *zero matrix*. We denote a zero matrix by the symbol [0].

With Definitions 6.8 and 6.9, it is immediate that a zero matrix is the unique additive identity to any $m \times n$ matrix. Thus, we state the following theorem.

theorem 6.2

For any $m \times n$ matrix **A,** there is a unique $m \times n$ matrix [0], such that **A** + [0] = [0] + **A = A.**

Now that we have an additive identity matrix for matrices of any order, we can easily prove that every matrix has a unique additive inverse.

theorem 6.3

For any $m \times n$ matrix **A,** there is a unique matrix −**A** of the same order such that −**A** + **A** = **A** + −**A** = [0]. Every entry of −**A** is the additive inverse of the corresponding entry of **A.**

Example

$$\begin{bmatrix} -3 & 1 & 5 \\ 2/3 & -5/8 & -4 \\ 0 & 5 & -1 \end{bmatrix} + \begin{bmatrix} 3 & -1 & -5 \\ -2/3 & 5/8 & 4 \\ 0 & -5 & 1 \end{bmatrix}$$

$$= \begin{bmatrix} (-3) + 3 & 1 + (-1) & 5 + (-5) \\ 2/3 + (-2/3) & -5/8 + 5/8 & -4 + 4 \\ 0 + 0 & 5 + (-5) & -1 + 1 \end{bmatrix} = \begin{bmatrix} 0 & 0 & 0 \\ 0 & 0 & 0 \\ 0 & 0 & 0 \end{bmatrix}.$$

We define subtraction of matrices to have exactly the same relationship to addition of matrices as subtraction of real numbers has to addition of real numbers. With this definition, we can prove a theorem which corresponds to the real-number theorem that states $a - b = a + (-b)$.

definition 6.10 (subtraction of matrices)
>For any $m \times n$ matrices **A**, **B**, and **C**, $\mathbf{A} - \mathbf{B} = \mathbf{C}$ if and only if $\mathbf{A} = \mathbf{B} + \mathbf{C}$.

theorem 6.4
>For any $m \times n$ matrices **A** and **B**, $\mathbf{A} - \mathbf{B} = \mathbf{A} + (-\mathbf{B})$.

>*proof*
>$\mathbf{A} - \mathbf{B} = \mathbf{C}$ implies that $\mathbf{A} = \mathbf{B} + \mathbf{C}$ by definition of subtraction.
>
>| | |
>|---|---|
>| $\mathbf{A} = \mathbf{C} + \mathbf{B}$ | by Theorem 6.1, Part (ii) |
>| $\mathbf{A} + (-\mathbf{B}) = (\mathbf{C} + \mathbf{B}) + (-\mathbf{B})$ | by Theorem 6.1, Part (iv) |
>| $\mathbf{A} + (-\mathbf{B}) = \mathbf{C} + (\mathbf{B} + (-\mathbf{B}))$ | by Theorem 6.1, Part (iii) |
>| $\mathbf{A} + (-\mathbf{B}) = \mathbf{C} + [0]$ | by Theorem 6.3 |
>| $\mathbf{A} + (-\mathbf{B}) = \mathbf{C}$ | by Theorem 6.2 |
>
>Therefore, since $\mathbf{A} - \mathbf{B} = \mathbf{C}$ and $\mathbf{A} + (-\mathbf{B}) = \mathbf{C}$, $\mathbf{A} - \mathbf{B} = \mathbf{A} + (-\mathbf{B})$.

Example

$$\begin{bmatrix} 5 & -3 & 2 \\ 4 & 1 & -6 \end{bmatrix} - \begin{bmatrix} 4 & 1 & 0 \\ -5 & 6 & 4 \end{bmatrix} = \begin{bmatrix} 5 & -3 & 2 \\ 4 & 1 & -6 \end{bmatrix} + \begin{bmatrix} -4 & -1 & 0 \\ 5 & -6 & -4 \end{bmatrix}$$

$$= \begin{bmatrix} 1 & -4 & 2 \\ 9 & -5 & -10 \end{bmatrix}.$$

We defined an operation of multiplication of a vector by a scalar. We now extend this definition to include all matrices instead of just vectors.

definition 6.11 (scalar multiplication)
>The product of any $m \times n$ matrix **A** and any real number c is the $m \times n$ matrix $c\mathbf{A}$ with entries ca_{ij} corresponding to every a_{ij} in **A**.

Example
(i) If

$$\mathbf{A} = \begin{bmatrix} -1 & 4 & 3 \\ 5 & -2 & 7 \end{bmatrix}$$

then

$$3\mathbf{A} = \begin{bmatrix} 3(-1) & 3(4) & 3(3) \\ 3(5) & 3(-2) & 3(7) \end{bmatrix} = \begin{bmatrix} -3 & 12 & 9 \\ 15 & -6 & 21 \end{bmatrix}.$$

(ii) If

$$\mathbf{B} = \begin{bmatrix} 4 & 2 \\ -1 & 3 \\ 5 & 7 \end{bmatrix}$$

then

$$-2\mathbf{B} = \begin{bmatrix} -2(4) & -2(2) \\ -2(-1) & -2(3) \\ -2(5) & -2(7) \end{bmatrix} = \begin{bmatrix} -8 & -4 \\ 2 & -6 \\ -10 & -14 \end{bmatrix}.$$

(iii) If

$$\mathbf{A} = \begin{bmatrix} 5 & 1 \\ -4 & 2 \end{bmatrix} \quad \text{and} \quad \mathbf{B} = \begin{bmatrix} -3 & 0 \\ 7 & -6 \end{bmatrix}$$

then

$$2\mathbf{A} + 3\mathbf{B} = \begin{bmatrix} 10 & 2 \\ -8 & 4 \end{bmatrix} + \begin{bmatrix} -9 & 0 \\ 21 & -18 \end{bmatrix} = \begin{bmatrix} 1 & 2 \\ 13 & -14 \end{bmatrix}.$$

Exercise Set 6.2

1. Find the transpose of each of the following matrices.

(a) $\begin{bmatrix} -2 & 3 \\ 5 & 1 \end{bmatrix}$

(f) $\begin{bmatrix} -2 & -3 & 5 \\ 7 & 11 & -13 \\ -17 & 19 & 23 \end{bmatrix}$

(b) $\begin{bmatrix} 4 & -2 \\ 6 & 13 \end{bmatrix}$

(g) $\begin{bmatrix} 2/3 & 4 & \pi \end{bmatrix}$

(c) $\begin{bmatrix} -5 & 1 & 3 \\ 2 & -3 & 2 \end{bmatrix}$

(h) $\begin{bmatrix} 5 & 1 \\ -4 & 3 \\ 2/3 & -5/8 \end{bmatrix}$

(d) $\begin{bmatrix} -15 & 2 & -4 \\ 5 & -2 & 3 \end{bmatrix}$

(i) $\begin{bmatrix} 7 \\ 11 \\ 2 \\ 4 \end{bmatrix}$

(e) $\begin{bmatrix} 7 & 11 & 13 \\ 4 & 8 & 17 \\ 9 & 6 & -3 \end{bmatrix}$

(j) $\begin{bmatrix} 7/8 & \sqrt{2} \\ \pi & 17 \\ -3 & 4 \\ 4/5 & 5/8 \end{bmatrix}$

2. Where

$$A = \begin{bmatrix} -2 & 1 & 3 \\ 4 & -3 & 6 \end{bmatrix}, \quad B = \begin{bmatrix} 4 & 0 & -4 \\ -2 & 4 & 7 \end{bmatrix}, \quad C = \begin{bmatrix} -4 & 1 & 6 \\ 7 & -5 & 0 \end{bmatrix},$$

find the following.

(a) $A + B$
(b) $A - B$
(c) $2A + 3B$
(d) $3B - 4C$
(e) $2A + 3B - 4C$
(f) $2(A + C)$
(g) $3A + 3C - B$
(h) $2A - 3B + 5C$

(i) $(-A) + (-B) + 3(-C)$
(j) $A^t + (-B)^t$
(k) $A + A^t$
(l) $B + C^t$
(m) $(B + C)^t$
(n) $B^t + C^t$
(o) $5(A + 3B)$

3. Find the matrix x that will make each statement true.

(a) $x + \begin{bmatrix} 15 & -1 \\ 4 & -3 \end{bmatrix} = \begin{bmatrix} 7 & 12 \\ 1 & 3 \end{bmatrix}$

(b) $x - \begin{bmatrix} 5 & 10 & 3 \\ -2 & 0 & 6 \end{bmatrix} = \begin{bmatrix} -3 & 4 & 1 \\ 5 & 0 & 6 \end{bmatrix}$

(c) $\begin{bmatrix} 5 & -8 & 8 & 3 \\ 3 & 4 & -6 & -7 \\ 8 & 2 & 4 & 2 \end{bmatrix} - x = \begin{bmatrix} -1 & 3 & 2 & 4 \\ 6 & 5 & 6 & -2 \\ -1 & 0 & 12 & 10 \end{bmatrix}$

(d) $\begin{bmatrix} -2 & 3 & 1 \\ 4 & 0 & 3 \\ 2 & 1 & 6 \end{bmatrix} + x = \begin{bmatrix} 4 & 10 & 5 \\ 12 & -1 & 2 \\ 4 & 3 & 6 \end{bmatrix}$

(e) $\begin{bmatrix} -1 & 2 & 10 & 3 \\ 0 & 13 & -4 & 20 \\ 8 & 11 & -8 & -5 \end{bmatrix} - x = \begin{bmatrix} 1 & 10 & 10 & 7 \\ -1 & 0 & 0 & 1 \\ 12 & 0 & 1 & 8 \end{bmatrix}$

4. Prove the following for all 3×3 matrices A, B, and C and scalars a, b, and c.

(a) $A^t + B^t = (A + B)^t$
(b) $c(A + B) = cA + cB$
(c) $(A + B) + C = A + (B + C)$
(d) $(A + B) = (B + A)$

(e) $(a + b)A = aA = bA$
(f) $A = (0) = A$
(g) $A + (-A) = (0)$

6.4 MATRIX MULTIPLICATION

We have defined the dot product of vectors, and this dot product is used to find the product of matrices that are not vectors.

The dot product of two vectors $\alpha = [a_1, a_2, a_3]$ and $\beta = [b_1, b_2, b_3]$ has been defined to be $a_1 b_1 + a_2 b_2 + a_3 b_3$. To make this definition consistent with matrix multiplication when α and β are considered matrices, we must consider β to be a column matrix

$$\beta = \begin{bmatrix} b_1 \\ b_2 \\ b_3 \end{bmatrix}.$$

Thus, we have

$$\alpha \cdot \beta = [a_1, a_2, a_3] \cdot \begin{bmatrix} b_1 \\ b_2 \\ b_3 \end{bmatrix} = a_1 b_1 + a_2 b_2 + a_3 b_3.$$

definition 6.12 (matrix multiplication)

For any $m \times n$ matrix $\mathbf{A}$ and $n \times r$ matrix $\mathbf{B}$, the product of $\mathbf{A}$ and $\mathbf{B}$ is the $m \times r$ matrix $\mathbf{AB}$, each of whose entries c_{ij} is the dot product of the ith row of $\mathbf{A}$ with the jth column of $\mathbf{B}$. If $\mathbf{B}$ does not have the same number of rows as $\mathbf{A}$ has columns, then $\mathbf{AB}$ is not defined.

Example

(i) If

$$\mathbf{A} = \begin{bmatrix} 5 & 1 & 4 \\ 2 & 0 & -3 \end{bmatrix} \quad \text{and} \quad \mathbf{B} = \begin{bmatrix} 6 & 4 \\ 2 & 1 \\ 4 & 0 \end{bmatrix}$$

then

$$\mathbf{AB} = \begin{bmatrix} [5,1,4] \cdot \begin{bmatrix} 6 \\ 2 \\ 4 \end{bmatrix} & [5,1,4] \cdot \begin{bmatrix} 4 \\ 1 \\ 0 \end{bmatrix} \\ [2,0,-3] \cdot \begin{bmatrix} 6 \\ 2 \\ 4 \end{bmatrix} & [2,0,-3] \cdot \begin{bmatrix} 4 \\ 1 \\ 0 \end{bmatrix} \end{bmatrix}$$

$$= \begin{bmatrix} 30 + 2 + 16 & 20 + 1 + 0 \\ 12 + 0 - 12 & 8 + 0 + 0 \end{bmatrix}$$

$$= \begin{bmatrix} 48 & 21 \\ 0 & 8 \end{bmatrix}.$$

(ii) If

$$A = \begin{bmatrix} 4 & 1 \\ -1 & 2 \end{bmatrix} \quad \text{and} \quad B = \begin{bmatrix} 5 & 2 & 3 \\ 4 & -1 & 1 \end{bmatrix}$$

then

$$AB = \begin{bmatrix} [4,1] \cdot \begin{bmatrix} 5 \\ 4 \end{bmatrix} & [4,1] \cdot \begin{bmatrix} 2 \\ -1 \end{bmatrix} & [4,1] \cdot \begin{bmatrix} 3 \\ 1 \end{bmatrix} \\ [-1,2] \cdot \begin{bmatrix} 5 \\ 4 \end{bmatrix} & [-1,2] \cdot \begin{bmatrix} 2 \\ -1 \end{bmatrix} & [-1,2] \cdot \begin{bmatrix} 3 \\ 1 \end{bmatrix} \end{bmatrix}$$

$$= \begin{bmatrix} 20+4 & 8-1 & 12+1 \\ -5+8 & -2-2 & -3+2 \end{bmatrix} = \begin{bmatrix} 24 & 7 & 13 \\ 3 & -4 & -1 \end{bmatrix}.$$

(iii) If

$$A = \begin{bmatrix} 6 & 4 \\ 2 & 1 \\ 4 & 0 \end{bmatrix} \quad \text{and} \quad B = \begin{bmatrix} 5 & -3 \\ 2 & 4 \\ -5 & 11 \end{bmatrix}$$

then **AB** is undefined, because **A** has 2 rows while **B** has 3 columns; that is,

$$[6,4] \cdot \begin{bmatrix} 5 \\ 2 \\ -5 \end{bmatrix}$$

is undefined.

Many properties are associated with scalar multiplication, but rather than state these properties as theorems, we have included them as exercises.

We found that the sum of two matrices of the same order is found by merely adding corresponding entries of the matrices. It may seem logical to define products in much the same way, but a product defined in such a way does not prove to be useful to us. The product as we have defined it proves to be more difficult, but also becomes extremely useful to both applied and theoretical problems.

The following theorems indicate real-number properties which are also valid for matrices. First note that **AB** can be defined while **BA** cannot, as in Part (i) of the previous example. Therefore, matrix multiplication is *not a commutative operation*.

Because of the extended and complicated notation involved in the proofs, the following theorems are stated without proof. Note that we state the distributive property both on the left and on the right. This is necessary, since we have no commutative property of multiplication.

theorem 6.5 (associativity)
> For any matrices **A, B,** and **C** for which **(AB)C** is defined, **(AB)C** = **A(BC)** (associative property of multiplication).

theorem 6.6 (distributivity)
> For any matrices **A, B,** and **C,**
> (i) if **A(B + C)** is defined, then **A(B + C)** = **AB + AC**   (left distributive property of multiplication over addition);
> (ii) if **(A + B)C** is defined, then **(A + B)C** = **AC + BC** (right distributive property of multiplication over addition).

definition 6.13
> I_n denotes a square matrix of order n, having only 1's for entries on the principal diagonal and 0's for all other entries.

Example

$$I_1 = [1], \quad I_2 = \begin{bmatrix} 1 & 0 \\ 0 & 1 \end{bmatrix}, \quad I_3 = \begin{bmatrix} 1 & 0 & 0 \\ 0 & 1 & 0 \\ 0 & 0 & 1 \end{bmatrix}.$$

theorem 6.7 (identity)
> For any $m \times n$ matrix **A,**
> (i) $I_m A = A$ (left identity);
> (ii) $A I_n = A$ (right identity).

Example

$$
\begin{bmatrix} 1 & 0 & 0 \\ 0 & 1 & 0 \\ 0 & 0 & 1 \end{bmatrix} \cdot \begin{bmatrix} 5 & -1 \\ 3 & 0 \\ -5 & 2 \end{bmatrix}
$$

$$
= \begin{bmatrix} 1(5) + 0(3) + 0(-5) & 1(-1) + 0(0) + 0(2) \\ 0(5) + 1(3) + 0(-5) & 0(-1) + 1(0) + 0(2) \\ 0(5) + 0(3) + 1(-5) & 0(-1) + 0(0) + 1(2) \end{bmatrix}
$$

$$
= \begin{bmatrix} 5 & -1 \\ 3 & 0 \\ -5 & 2 \end{bmatrix}
$$

$$
\begin{bmatrix} 5 & -1 \\ 3 & 0 \\ -5 & 2 \end{bmatrix} \cdot \begin{bmatrix} 1 & 0 \\ 0 & 1 \end{bmatrix} = \begin{bmatrix} 5(1) + (-1)(0) & 5(0) + (-1)(1) \\ 3(1) + 0(0) & 3(0) + 0(1) \\ -5(1) + 2(0) & -5(0) + 2(1) \end{bmatrix}
$$

$$
= \begin{bmatrix} 5 & -1 \\ 3 & 0 \\ -5 & 2 \end{bmatrix}.
$$

An additional property of multiplication of matrices, which is not treated in this section, is the multiplicative inverse property. This is an important property — in fact, the most important of this chapter to the solving of systems of equations. Finding the multiplicative inverse of a matrix (when it exists), however, requires an understanding of concepts and skills that we have not yet discussed. Thus, we leave this important topic until later in the chapter when we devote a complete section to it.

Exercise Set 6.3

1. Find the following products.

(a) $\begin{bmatrix} 5 & 1 & 6 \end{bmatrix} \begin{bmatrix} 4 \\ 2 \\ 3 \end{bmatrix}$

(b) $\begin{bmatrix} 4 & 2 \end{bmatrix} \begin{bmatrix} 8 \\ -3 \end{bmatrix}$

(c) $\begin{bmatrix} 9 & 3 \end{bmatrix} \begin{bmatrix} 5 & 1 \\ 2 & 3 \end{bmatrix}$

(d) $\begin{bmatrix} 7 & 2 \end{bmatrix} \begin{bmatrix} 6 & 8 & 3 \\ 5 & 1 & 4 \end{bmatrix}$

(e) $\begin{bmatrix} 4 & -3 \\ 1 & 4 \end{bmatrix}\begin{bmatrix} 3 \\ 2 \end{bmatrix}$

(f) $\begin{bmatrix} 2 & 5 \\ 1 & 9 \end{bmatrix}\begin{bmatrix} 4 & 1 & -1 \\ 3 & 6 & 2 \end{bmatrix}$

(g) $\begin{bmatrix} 2 & 1 & 3 \\ 0 & 3 & 1 \\ 5 & 0 & 2 \end{bmatrix}\begin{bmatrix} 0 \\ -1 \\ 2 \end{bmatrix}$

(h) $\begin{bmatrix} -1 & 2 & 3 \\ 5 & 4 & 3 \\ 2 & 6 & 1 \end{bmatrix}\begin{bmatrix} 1 & 6 \\ 4 & 3 \\ 5 & 2 \end{bmatrix}$

(i) $\begin{bmatrix} 5 & 2 \\ 1 & 0 \\ 3 & 8 \end{bmatrix}\begin{bmatrix} 4 & 1 & -3 & 6 & 2 \\ 5 & -8 & 1 & -9 & 3 \end{bmatrix}$

(j) $\begin{bmatrix} 4 & 1 & 5 & 8 & 7 \\ 2 & 1 & -1 & 3 & 4 \end{bmatrix}\begin{bmatrix} 1 & 4 \\ 0 & 1 \\ 3 & 5 \\ 1 & 0 \\ 2 & -2 \end{bmatrix}$

(k) $\begin{bmatrix} -5 & 1 & -3 \end{bmatrix}\begin{bmatrix} 4 & -1 & 2 \\ 3 & 2 & -6 \\ 3 & 1 & 4 \end{bmatrix}$

(l) $\begin{bmatrix} 2 & 0 & 0 \\ 0 & 2 & 0 \\ 0 & 0 & 2 \end{bmatrix}\begin{bmatrix} 5 & 1 & 3 \\ 0 & 4 & 8 \\ 9 & -6 & 12 \end{bmatrix}$

2. Show that matrix multiplication is not commutative by a counter-example with two 2×2 matrices.

3. If A is a 9×3 matrix and B is a 3×12 matrix, what is the order of AB?

4. If A is a 5×201 matrix and B is a 201×2 matrix, what is the order of AB?

5. If A and B are 2×2 matrices, show that the following are not true.
   (a) $(A + B)^2 = A^2 + 2AB + B^2$
   (b) $(A - B)(A + B) = A^2 - B^2$

6. Prove the following for any 2×2 matrices A, B, and C and real numbers a, b, and c.
   (a) $(AB)^t = B^t A^t$
   (b) $(0)A = (0)$
   (c) $I_3 A = A I_3 = A$

(d) $(ab)\mathbf{A} = a(b\mathbf{A})$

(e) $(b + c)\mathbf{A} = b\mathbf{A} + c\mathbf{A}$

(f) $c(0) = c$

(g) $(\mathbf{A}^t)^t = \mathbf{A}$

6.5 DETERMINANTS

Associated with each square matrix of real numbers is a real number called the *determinant* of the matrix. Just as the magnitude of a vector is not the vector itself, the determinant of a matrix is not the rectangular array, but merely a real number associated with it. Thus, the term "determinant" has meaning only when associated with a square matrix. Before defining the determinant of a matrix of order n, we define the determinant of a matrix of order 2.

definition 6.14 (determinant of a 2 × 2 matrix)

If $\mathbf{A}$ is the 2×2 matrix

$$\begin{bmatrix} a_{11} & a_{12} \\ a_{21} & a_{22} \end{bmatrix}$$

of real numbers, then we define the real number $a_{11} \cdot a_{22} - a_{21} \cdot a_{12}$ to be the determinant of $\mathbf{A}$, denoted $|\mathbf{A}|$. We write

$$|\mathbf{A}| = \begin{vmatrix} a_{11} & a_{12} \\ a_{21} & a_{22} \end{vmatrix} = (a_{11})(a_{22}) - (a_{21})(a_{12}).$$

Example

(i) If

$$\mathbf{A} = \begin{bmatrix} 3 & -2 \\ 4 & 5 \end{bmatrix}$$

then

$$|\mathbf{A}| = \begin{vmatrix} 3 & -2 \\ 4 & 5 \end{vmatrix} = 3(5) - 4(-2) = 23.$$

(ii) If

$$A = \begin{bmatrix} 3 & 1 \\ -6 & 2 \end{bmatrix}$$

then

$$|A| = \begin{vmatrix} 3 & 1 \\ -6 & 2 \end{vmatrix} = 3(2) - (-6)(1) = 12.$$

definition 6.15 (cofactor)

With each entry a_{ij} of a square matrix of order n is associated a *cofactor*, which is the product of $(-1)^{i+j}$ and the determinant of the matrix remaining after eliminating the ith row and the jth column.

Example

Given the matrix

$$A = \begin{bmatrix} 5 & -2 & 1 \\ 3 & 0 & 4 \\ -7 & 1 & 2 \end{bmatrix}.$$

The cofactor of 5 is

$$(-1)^2 \cdot \begin{vmatrix} 0 & 4 \\ 1 & 2 \end{vmatrix} = 1(0 - 4) = -4.$$

The cofactor of 3 is

$$(-1)^3 \cdot \begin{vmatrix} -2 & 1 \\ 1 & 2 \end{vmatrix} = -1(-4 - 1) = 5.$$

The cofactor of -7 is

$$(-1)^4 \cdot \begin{vmatrix} -2 & 1 \\ 0 & 4 \end{vmatrix} = 1(-8 - 0) = -8.$$

The cofactor of 0 is

$$(-1)^4 \cdot \begin{vmatrix} 5 & 1 \\ -7 & 2 \end{vmatrix} = 1(10 - (-7)) = 1(17) = 17.$$

The cofactor of 1 (in the second column) is

$$(-1)^5 \cdot \begin{vmatrix} 5 & 1 \\ 3 & 4 \end{vmatrix} = -1(20 - 3) = -17.$$

The cofactor of 1 (in the third column) is

$$(-1)^4 \cdot \begin{vmatrix} 3 & 0 \\ -7 & 1 \end{vmatrix} = 1(3 - 0) = 3.$$

The cofactor of 2 is

$$(-1)^6 \cdot \begin{vmatrix} 5 & -2 \\ 3 & 0 \end{vmatrix} = 1(0 + 6) = 6.$$

We are now prepared to define the determinant of an $n \times n$ matrix when $n > 2$. The only additional case not yet considered is the 1×1 matrix, whose determinant is its only entry.

definition 6.16 (determinant of an $n \times n$ matrix)
    The determinant of any $n \times n$ matrix for which $n > 2$ is the dot product of any row or column vector with the vector of its corresponding cofactors.

Definitions 6.15 and 6.16 may, at first glance, seem to be circular, since a cofactor is defined in terms of a determinant and a determinant is defined in terms of a cofactor. The definitions are not circular, however. The determinant of a 2×2 matrix is defined independently of a cofactor. The cofactor of each entry of a 3×3 matrix is defined in terms of determinants of 2×2 matrices. Thus, the determinant of a 4×4 matrix is defined in terms of cofactors of entries of a 3×3 matrix, allowing the cofactors of the entries of a 4×4 matrix to be defined, and so on, in like manner to any $n \times n$ matrix.

Example
(i)   If

$$\mathbf{A} = \begin{bmatrix} 5 & -3 & 2 \\ 4 & 1 & 6 \\ -3 & 2 & 4 \end{bmatrix}$$

then we can find its determinant, using any row or column.

Using the first column, we obtain

$$|A| = \begin{vmatrix} 5 & -3 & 2 \\ 4 & 1 & 6 \\ -3 & 2 & 4 \end{vmatrix}$$

$$= 5\begin{vmatrix} 1 & 6 \\ 2 & 4 \end{vmatrix} - 4\begin{vmatrix} -3 & 2 \\ 2 & 4 \end{vmatrix} - 3\begin{vmatrix} -3 & 2 \\ 1 & 6 \end{vmatrix}$$

$$= 5(4 - 12) - 4(-12 - 4) - 3(-18 - 2)$$
$$= 5(-8) - 4(-16) - 3(-20)$$
$$= -40 + 64 + 60$$
$$= 84.$$

We can find the same determinant, using the third row.

$$|A| = \begin{vmatrix} 5 & -3 & 2 \\ 4 & 1 & 6 \\ -3 & 2 & 4 \end{vmatrix}$$

$$= -3\begin{vmatrix} -3 & 2 \\ 1 & 6 \end{vmatrix} - 2\begin{vmatrix} 5 & 2 \\ 4 & 6 \end{vmatrix} + 4\begin{vmatrix} 5 & -3 \\ 4 & 1 \end{vmatrix}$$

$$= -3(-18 - 2) - 2(30 - 8) + 4(5 + 12)$$
$$= -3(-20) - 2(22) + 4(17)$$
$$= +60 - 44 + 68$$
$$= 84.$$

(ii) If

$$B = \begin{bmatrix} 5 & 1 & 4 & 3 \\ -2 & 0 & 1 & 0 \\ 3 & 0 & -1 & 2 \\ 2 & 4 & 3 & 5 \end{bmatrix}$$

then we can find $|B|$ most easily by selecting either the second row or the second column, since both contain two zeros.

$$|B| = \begin{vmatrix} 5 & 1 & 4 & 3 \\ -2 & 0 & 1 & 6 \\ 3 & 1 & -1 & 2 \\ 2 & 4 & 3 & 5 \end{vmatrix}$$

$$|\mathbf{B}| = -1 \begin{vmatrix} 2 & 1 & 0 \\ 3 & -1 & 2 \\ 2 & 3 & 5 \end{vmatrix} + 0 \begin{vmatrix} 5 & 4 & 3 \\ 3 & -1 & 2 \\ 2 & 3 & 5 \end{vmatrix}$$

$$+ 0 \begin{vmatrix} 5 & 4 & 3 \\ 2 & 1 & 0 \\ 2 & 3 & 5 \end{vmatrix} + 4 \begin{vmatrix} 5 & 4 & 3 \\ 2 & 1 & 0 \\ 3 & -1 & 2 \end{vmatrix}$$

$$= -1 \begin{vmatrix} 2 & 1 & 0 \\ 3 & -1 & 2 \\ 2 & 3 & 5 \end{vmatrix} + 4 \begin{vmatrix} 5 & 4 & 3 \\ 2 & 1 & 0 \\ 3 & -1 & 2 \end{vmatrix}$$

$$= -1 \left(2 \begin{vmatrix} -1 & 2 \\ 3 & 5 \end{vmatrix} - 1 \begin{vmatrix} 3 & 2 \\ 2 & 5 \end{vmatrix} + 0 \begin{vmatrix} 3 & -1 \\ 2 & 3 \end{vmatrix} \right)$$

$$+ 4 \left(-2 \begin{vmatrix} 4 & 3 \\ -1 & 2 \end{vmatrix} + 1 \begin{vmatrix} 5 & 3 \\ 3 & 2 \end{vmatrix} - 0 \begin{vmatrix} 5 & 4 \\ 3 & -1 \end{vmatrix} \right)$$

$$= -1(2(-11) - 1(11)) + 4(-2(11) + 1(1))$$
$$= -1(-22 - 11) + 4(-22 + 1)$$
$$= 33 - 84$$
$$= -51.$$

Exercise Set 6.4

1. Find the cofactor of each entry of the matrix

$$\begin{bmatrix} 3 & 1 & 4 \\ 2 & 0 & 6 \\ -5 & 3 & 1 \end{bmatrix}$$

2. Find the determinant of the matrix in Problem 1:
   (a) using the first row and its cofactors;
   (b) using the second column and its cofactors;
   (c) using the third row and its cofactors.

3. Find the determinant of each of the following matrices.

   (a) $[2 \quad 3]$

   (b) $\begin{bmatrix} -2 & 3 \\ 4 & 7 \end{bmatrix}$

   (c) $\begin{bmatrix} 5 & 0 \\ 1 & 6 \end{bmatrix}$

   (d) $\begin{bmatrix} 1 & 0 \\ 5 & 0 \end{bmatrix}$

   (e) $\begin{bmatrix} 5 & 1 & 2 \\ 3 & 3 & 1 \\ 1 & 3 & 6 \end{bmatrix}$

   (f) $\begin{bmatrix} 2 & 0 & 2 \\ 3 & 1 & 6 \\ 4 & 1 & 3 \end{bmatrix}$

(g) $\begin{bmatrix} 0 & 2 & 1 \\ 3 & 0 & 3 \\ 1 & 2 & 0 \end{bmatrix}$ (i) $\begin{bmatrix} 4 & 2 & -3 \\ -6 & 5 & 4 \\ 7 & -6 & 5 \end{bmatrix}$

(h) $\begin{bmatrix} 5 & 0 & 0 \\ 2 & 1 & 3 \\ 7 & -3 & 4 \end{bmatrix}$ (j) $\begin{bmatrix} 5 & -1 & 3 & 2 \\ 2 & 1 & 0 & 3 \\ 4 & 0 & 2 & 2 \\ -3 & 4 & 0 & -6 \end{bmatrix}$

4. Solve the following for x.

(a) $\begin{vmatrix} 5 & 2 & -3 \\ 0 & 11 & 7 \\ x & 0 & 0 \end{vmatrix} = 5$ (d) $\begin{vmatrix} x & 2 & 0 \\ 5 & 4 & 2 \\ -3 & 6 & 1 \end{vmatrix} = -10$

(b) $\begin{vmatrix} x & 3 \\ 2 & 1 \end{vmatrix} = 4$ (e) $\begin{vmatrix} 5 & 8 & 3 \\ 0 & x & 2 \\ 4 & -3 & 1 \end{vmatrix} = 7$

(c) $\begin{vmatrix} 4 & 3 & x \\ 5 & 0 & 3 \\ 1 & 2 & -4 \end{vmatrix} = 3$ (f) $\begin{vmatrix} x^2 & x & 1 \\ 3 & 1 & 2 \\ 4 & 2 & 5 \end{vmatrix} = -10$

5.
$$\mathbf{A} = \begin{bmatrix} 3 & 1 & 2 \\ -1 & 5 & 1 \\ 2 & 0 & 5 \end{bmatrix}, \mathbf{B} = \begin{bmatrix} -1 & 5 & 1 \\ 3 & 1 & 2 \\ 2 & 0 & 5 \end{bmatrix}$$ (A with the first two rows inter-changed)

$$\mathbf{C} = \begin{bmatrix} 3 & 1 & 2 \\ -1 & 5 & 1 \\ 1 & 5 & 6 \end{bmatrix}$$ (A with the third row replaced by the sum of the second and third rows)

$$\mathbf{D} = \begin{bmatrix} 3 & 1 & 2 \\ -2 & 10 & 2 \\ 2 & 0 & 5 \end{bmatrix}$$ (A with all entries in the second row multiplied by 2)

Find the following.

(a) $|\mathbf{A}|$ (g) $2|\mathbf{A}| + 2|\mathbf{B}|$
(b) $|\mathbf{B}|$ (h) $2|\mathbf{A} + \mathbf{B}|$
(c) $|\mathbf{C}|$ (i) $|\mathbf{BC}|$
(d) $|\mathbf{D}|$ (j) $|\mathbf{B}| \cdot |\mathbf{C}|$
(e) $|\mathbf{A} + \mathbf{B}|$ (k) $|2\mathbf{A}|$
(f) $|\mathbf{A}| + |\mathbf{B}|$ (l) $2|\mathbf{A}|$

6. Give an example of each of the following, which can be proved as theorems for determinants.

(a) If each entry of any row (or column) of a square matrix **A** is 0, then the determinant of **A** is equal to 0.

(b) If any two rows (or columns) of a square matrix **A** are interchanged, then the determinant of the resulting matrix is $-|\mathbf{A}|$.

(c) If any two rows (or columns) of a square matrix **A** have corresponding entries that are equal, then $|\mathbf{A}| = 0$.

(d) If we multiply each entry of any row (or column) of a square matrix **A** by a constant k, then the determinant of the resulting matrix is $k|\mathbf{A}|$.

(e) If every entry of one row (or column) of a square matrix **A** is multiplied by the same real number k and each product is then added to the corresponding entry in another row (or column), the determinant of the resulting matrix is equal to $k|\mathbf{A}|$.

(f) If the corresponding entries of two rows (or columns) of a square matrix **A** are in the same ratio, then $|\mathbf{A}| = 0$.

6.6 THE INVERSE OF A MATRIX

We have found that the mathematical system consisting of the set of all $m \times n$ matrices and the operations of addition and multiplication of matrices has many, but not all, of the properties of real numbers. One property, which is often present with certain matrices, has not yet been discussed. This is the *multiplicative inverse* or, more simply, *inverse* property (when we refer to the inverse of a matrix, unless otherwise specified, we mean the multiplicative inverse).

Consider the multiplicative inverse property of real numbers, which states that $a^{-1} \cdot a = a \cdot a^{-1} = 1$. If a multiplicative inverse property of matrices is to correspond to this, then we are restricted to the consideration of matrices **A** and $\mathbf{A}^{-1}$ such that both $\mathbf{A} \cdot \mathbf{A}^{-1}$ and $\mathbf{A}^{-1} \cdot \mathbf{A}$ are defined. The only time the product of two matrices **A** and **B** is defined both for **AB** and for **BA** is when **A** and **B** are square matrices of the same order.

definition 6.17 (inverse of a square matrix)
If **A** is a square matrix of order n and there exists a square matrix **B** of order n such that $\mathbf{AB} = \mathbf{BA} = \mathbf{I}_n$, then **B** is said to be the inverse of **A**, denoted $\mathbf{B} = \mathbf{A}^{-1}$. If $\mathbf{A}^{-1}$ exists, then **A** is said to be a *nonsingular* matrix. If $\mathbf{A}^{-1}$ does not exist, then **A** is said to be *singular*. (*Note:* $\mathbf{A}^{-1}$ does not mean $1/\mathbf{A}$.)

Example

If

$$\mathbf{A} = \begin{bmatrix} -2 & 1 \\ 4 & -3 \end{bmatrix} \quad \text{and} \quad \mathbf{B} = \begin{bmatrix} -3/2 & -1/2 \\ -2 & -1 \end{bmatrix}$$

then

$$\mathbf{AB} = \begin{bmatrix} -2 & 1 \\ 4 & -3 \end{bmatrix} \begin{bmatrix} -3/2 & -1/2 \\ -2 & -1 \end{bmatrix} = \begin{bmatrix} 3-2 & 1-1 \\ -6+6 & -2+3 \end{bmatrix} = \begin{bmatrix} 1 & 0 \\ 0 & 1 \end{bmatrix}$$

and

$$\mathbf{BA} = \begin{bmatrix} -3/2 & -1/2 \\ -2 & -1 \end{bmatrix} \begin{bmatrix} -2 & 1 \\ 4 & -3 \end{bmatrix}$$

$$= \begin{bmatrix} 3-2 & -3/2+3/2 \\ 4-4 & -2+3 \end{bmatrix} = \begin{bmatrix} 1 & 0 \\ 0 & 1 \end{bmatrix}.$$

Therefore, $\mathbf{B} = \mathbf{A}^{-1}$ and $\mathbf{A} = \mathbf{B}^{-1}$.

Not all square matrices have inverses, and even when it is known that an inverse of a particular matrix exists, it is not obvious what that inverse is or how to find it. There are many ways of finding the inverse of a matrix, and the derivation and proof of theorems dictating such methods could easily occupy a complete chapter. Our interest in matrices in this book, however, is limited to their application to problem solving. Thus, we give only the method that can most easily be applied to the type of problems that will be discussed. This method makes use of the concept of the *adjoint* of a matrix, which we now define.

definition 6.18 (adjoint)

Let **A** be a square matrix and **B** be the square matrix of the same order, each of whose entries is the cofactor of the corresponding entry of **A**. Then the matrix $\mathbf{B}^t$ is called the adjoint matrix of **A** and is denoted adj**A**.

Example

If

$$\mathbf{A} = \begin{bmatrix} 5 & -1 & 0 \\ 3 & 2 & -2 \\ 0 & 1 & -3 \end{bmatrix}$$

then

$$\text{adjA} = \begin{bmatrix} \begin{vmatrix} 2 & -2 \\ 1 & -3 \end{vmatrix} & -\begin{vmatrix} 3 & -2 \\ 0 & -3 \end{vmatrix} & \begin{vmatrix} 3 & 2 \\ 0 & 1 \end{vmatrix} \\ -\begin{vmatrix} -1 & 0 \\ 1 & -3 \end{vmatrix} & \begin{vmatrix} 5 & 0 \\ 0 & -3 \end{vmatrix} & -\begin{vmatrix} 5 & -1 \\ 0 & 1 \end{vmatrix} \\ \begin{vmatrix} -1 & 0 \\ 2 & -2 \end{vmatrix} & -\begin{vmatrix} 5 & 0 \\ 3 & -2 \end{vmatrix} & \begin{vmatrix} 5 & -1 \\ 3 & 2 \end{vmatrix} \end{bmatrix}^t$$

$$= \begin{bmatrix} -4 & 9 & 3 \\ -3 & -15 & -5 \\ 2 & 10 & 13 \end{bmatrix}^t = \begin{bmatrix} -4 & -3 & 2 \\ 9 & -15 & 10 \\ 3 & -5 & 13 \end{bmatrix}.$$

We are now ready to state a theorem that provides a method of determining the inverse of any square matrix, provided that the inverse exists.

theorem 6.8

For any square matrix **A** for which $\mathbf{A}^{-1}$ exists,

$$\mathbf{A}^{-1} = \frac{1}{|\mathbf{A}|}\,\text{adjA}.$$

Example

If

$$\mathbf{A} = \begin{bmatrix} 5 & -1 & 0 \\ 3 & 2 & -2 \\ 0 & 1 & -3 \end{bmatrix}$$

as in the previous example, then

$$\text{adjA} = \begin{bmatrix} -4 & -3 & 2 \\ 9 & -15 & 10 \\ 3 & -5 & 13 \end{bmatrix}$$

and $|\mathbf{A}| = 5(-4) + 7(9) + 0(3) = -29$, using the first row. Therefore,

$$\mathbf{A}^{-1} = \frac{1}{|\mathbf{A}|}\,(\text{adjA}) = \frac{-1}{29}\begin{bmatrix} -4 & -3 & 2 \\ 9 & -15 & 10 \\ 3 & -5 & 13 \end{bmatrix}.$$

We check the results by multiplication as follows:

$$\mathbf{A}^{-1} \cdot \mathbf{A} = -\frac{1}{29} \begin{bmatrix} -4 & -3 & 2 \\ 9 & -15 & 10 \\ 3 & -5 & 13 \end{bmatrix} \begin{bmatrix} 5 & -1 & 0 \\ 3 & 2 & -2 \\ 0 & 1 & -3 \end{bmatrix} = \begin{bmatrix} 1 & 0 & 0 \\ 0 & 1 & 0 \\ 0 & 0 & 1 \end{bmatrix}.$$

From Theorem 6.8 it is clear that if $|\mathbf{A}| = 0$, then $\mathbf{A}^{-1}$ does not exist. In fact, $\mathbf{A}^{-1}$ exists if and only if $|\mathbf{A}|$ exists and is not 0.

theorem 6.9

For any square matrix $\mathbf{A}$, $\mathbf{A}$ is nonsingular if and only if $|\mathbf{A}| \neq 0$.

Example

If

$$\mathbf{A} = \begin{bmatrix} -9 & 4 \\ 27 & -12 \end{bmatrix}$$

then $|\mathbf{A}| = 108 - 108 = 0$. Therefore, $\mathbf{A}^{-1}$ does not exist; so $\mathbf{A}$ is a singular matrix.

Thus, we have a method of determining whether or not a matrix has an inverse and a method of finding the inverse when it exists.

Exercise Set 6.5

1.   Identify each of the following matrices as singular or nonsingular.

(a) $\begin{bmatrix} -2 & 7 \\ -1 & 2 \end{bmatrix}$

(e) $\begin{bmatrix} 3 & 1 & 4 \\ -2 & 1 & 3 \\ 0 & 4 & 0 \end{bmatrix}$

(b) $\begin{bmatrix} 1 & 0 \\ 0 & 1 \end{bmatrix}$

(f) $\begin{bmatrix} -5 & 1 & 2 \\ 6 & 0 & -7 \\ 9 & 1 & 1 \end{bmatrix}$

(c) $\begin{bmatrix} -1 & 1 & -2 \\ 1 & 3 & 2 \\ 0 & 2 & 0 \end{bmatrix}$

(g) $\begin{bmatrix} -2 & 3 & 4 \\ 5 & 1 & -3 \\ -2 & 6 & 8 \end{bmatrix}$

(d) $\begin{bmatrix} 6 & -3 \\ -5 & -2 \end{bmatrix}$

(h) $\begin{bmatrix} 4 & 0 & 4 \\ -5 & 1 & 6 \\ 3 & -3 & 2 \end{bmatrix}$

(i) $\begin{bmatrix} 5 & -1 & 0 \\ -3 & 3 & 4 \\ 4 & 8 & 12 \end{bmatrix}$ (j) $\begin{bmatrix} 1 & 2 & -1 & 3 \\ 2 & 0 & 1 & 4 \\ 3 & 2 & 1 & 0 \\ 6 & 4 & 1 & 7 \end{bmatrix}$

2. Find the adjoint of each of the following matrices.

(a) $\begin{bmatrix} -3 & 2 \\ 0 & 5 \end{bmatrix}$ (f) $\begin{bmatrix} 0 & 4 & 2 \\ 1 & 0 & 2 \\ 0 & -1 & 1 \end{bmatrix}$

(b) $\begin{bmatrix} 6 & -3 \\ -5 & -2 \end{bmatrix}$ (g) $\begin{bmatrix} 4 & -7 & 16 \\ 3 & 1 & 10 \\ 5 & -6 & 10 \end{bmatrix}$

(c) $\begin{bmatrix} 2 & 3 \\ -2 & 6 \end{bmatrix}$ (h) $\begin{bmatrix} 3 & 0 & 8 \\ 3 & 0 & 0 \\ 6 & 3 & 2 \end{bmatrix}$

(d) $\begin{bmatrix} 4 & -5 \\ 9 & -8 \end{bmatrix}$ (i) $\begin{bmatrix} 2 & 3 & 0 \\ 3 & 1 & -2 \\ 4 & -2 & 3 \end{bmatrix}$

(e) $\begin{bmatrix} 4 & 1 & 3 \\ 2 & -2 & 2 \\ 1 & 0 & 0 \end{bmatrix}$ (j) $\begin{bmatrix} 5 & -4 & -1 \\ 8 & 9 & -6 \\ -2 & 5 & 3 \end{bmatrix}$

3. Find the inverse, if it exists, of each of the following matrices.

(a) $\begin{bmatrix} 1 & 0 \\ 0 & 1 \end{bmatrix}$ (f) $\begin{bmatrix} 4 & 1 & 3 \\ 2 & 0 & 4 \\ -1 & 0 & -2 \end{bmatrix}$

(b) $\begin{bmatrix} 0 & 0 \\ 0 & 0 \end{bmatrix}$ (g) $\begin{bmatrix} 7 & -2 & 6 \\ 4 & 1 & 2 \\ 0 & 5 & 1 \end{bmatrix}$

(c) $\begin{bmatrix} 5 & 7 \\ -3 & 2 \end{bmatrix}$ (h) $\begin{bmatrix} -2 & 1 & 1 \\ 4 & -3 & 7 \\ 4 & 2 & -3 \end{bmatrix}$

(d) $\begin{bmatrix} 0 & 1 \\ 1 & 0 \end{bmatrix}$ (i) $\begin{bmatrix} 3 & 1 & 3 \\ 2 & 0 & 4 \\ 5 & 0 & 0 \end{bmatrix}$

(e) $\begin{bmatrix} -7 & 3 \\ 21 & -9 \end{bmatrix}$ (j) $\begin{bmatrix} 1 & 2 & -1 & 3 \\ 2 & 0 & 1 & 4 \\ 3 & 2 & 1 & 0 \\ 6 & 4 & 1 & 7 \end{bmatrix}$

4. Show that for any 2×2 nonsingular matrices **A** and **B**, $|AB| =$

$|\mathbf{A}| \cdot |\mathbf{B}|$. (The same is true for $n \times n$. The proof is difficult in general.)

5. Show that for any nonsingular matrix $\mathbf{A}$, $(\mathbf{A}^t)^{-1} = (\mathbf{A}^{-1})^t$.
6. Show that for any nonsingular matrix $\mathbf{A}$, $|\mathbf{A}^{-1}| = 1/|\mathbf{A}|$.

6.7 SYSTEMS OF LINEAR EQUATIONS

One application of matrices is found in the solving of systems of equations. The advantage of using matrices for this purpose does not become evident until we begin solving systems of three or more equations. In fact, the labor involved in solving three linear equations in three variables is much the same when matrices are used as when they are not. The greater the number of equations in the system, the greater is the advantage of using matrices. However, most of the examples used here are of systems of only three equations, since examples of larger systems use exactly the same principle, but merely involve more extensive arithmetic.

One method of solving systems of linear equations with matrices makes use of *elementary row operations*. These are operations on the rows of a matrix, which include the following:

1) Interchange any two rows.
2) Multiply any row by a nonzero scalar.
3) Add a multiple of any row to any other row.

definition 6.19 (row equivalence)

Two $m \times n$ matrices $\mathbf{A}$ and $\mathbf{B}$ are said to be *row equivalent* if and only if $\mathbf{B}$ can be obtained from $\mathbf{A}$ by a finite number of elementary row operations.

Example

(i) If

$$\mathbf{A} = \begin{bmatrix} 1 & 5 & 3 \\ -2 & 1 & 6 \\ 4 & 1 & -3 \end{bmatrix} \quad \text{and} \quad \mathbf{B} = \begin{bmatrix} 1 & 5 & 3 \\ 4 & 1 & -3 \\ -2 & 1 & 6 \end{bmatrix}$$

then $\mathbf{A}$ and $\mathbf{B}$ are row equivalent, because $\mathbf{B}$ can be obtained from $\mathbf{A}$ by interchanging the second and third rows.

(ii) If

$$C = \begin{bmatrix} 5 & -1 & 2 \\ 1 & 0 & 3 \\ 2 & 5 & 7 \end{bmatrix} \quad \text{and} \quad D = \begin{bmatrix} 3 & 0 & -4 \\ 1 & 0 & 3 \\ 3 & 5 & 10 \end{bmatrix}$$

then C and D are row equivalent, because D can be obtained from C by adding -2 times the second row to the first and 1 times the second row to the third.

Through elementary row operations, any matrix can be reduced to a useful form, which we call the *reduced echelon* form of the matrix.

definition 6.20 (reduced echelon form)

A matrix is said to be in *reduced echelon* form if and only if the following conditions hold:

(a) All nonzero rows precede all zero rows (if there are any zero rows).

(b) Each nonzero row has more zeros preceding the first nonzero entry than the preceding row.

(c) The first nonzero entry in each row is 1.

(d) The first nonzero entry in each row is the only nonzero entry in its column.

Example

The matrices

$$\begin{bmatrix} 1 & 0 & 0 \\ 0 & 1 & 0 \\ 0 & 0 & 1 \end{bmatrix}, \quad \begin{bmatrix} 1 & 0 & 5 \\ 0 & 1 & 3 \end{bmatrix}, \quad \text{and} \quad \begin{bmatrix} 1 & 0 \\ 0 & 1 \\ 0 & 0 \\ 0 & 0 \end{bmatrix}$$

are in reduced echelon form, but the following are not:

$$\begin{bmatrix} 0 & 0 & 0 & 0 \\ 0 & 1 & 0 & 0 \\ 0 & 0 & 1 & 0 \\ 0 & 0 & 0 & 1 \end{bmatrix} \quad \begin{bmatrix} 1 & 2 \\ 0 & 1 \\ 0 & 0 \end{bmatrix}$$

$$\begin{bmatrix} 1 & 0 & 0 \\ 0 & 0 & 1 \\ 0 & 1 & 0 \end{bmatrix} \quad \begin{bmatrix} 1 & 0 \\ 0 & 3 \\ 0 & 0 \end{bmatrix}.$$

There is no unique sequence of elementary row operations that must be used to obtain the reduced echelon form of a matrix. In fact, 50 people may use 50 different sequences of operation and obtain the same reduced echelon form. Now, consider the system of linear equations

$$3x + 2y - z = 1$$
$$6x - y + 3z = -8$$
$$x + 3y + z = 16.$$

We can express this system as the matrix product

$$\begin{bmatrix} 3 & 2 & -1 \\ 6 & -1 & 3 \\ 1 & 3 & 1 \end{bmatrix} \begin{bmatrix} x \\ y \\ z \end{bmatrix} = \begin{bmatrix} 1 \\ -8 \\ 16 \end{bmatrix}.$$

We can now solve this system in either of two ways with tools that we have already developed. One method is to use elementary row operations on the left matrix to obtain its reduced echelon form, and at the same time perform the exact same elementary row operations on the right matrix. For example, we obtain

$$\begin{bmatrix} 4 & 5 & 0 \\ 6 & -1 & 3 \\ 1 & 3 & 1 \end{bmatrix} \begin{bmatrix} x \\ y \\ z \end{bmatrix} = \begin{bmatrix} 17 \\ -8 \\ 16 \end{bmatrix}$$ by adding the third row to the first;

$$\begin{bmatrix} 4 & 5 & 0 \\ 3 & -10 & 0 \\ 1 & 3 & 1 \end{bmatrix} \begin{bmatrix} x \\ y \\ z \end{bmatrix} = \begin{bmatrix} 17 \\ -56 \\ 16 \end{bmatrix}$$ by adding -3 times the third row to the second;

$$\begin{bmatrix} 4 & 5 & 0 \\ 11 & 0 & 0 \\ 1 & 3 & 1 \end{bmatrix} \begin{bmatrix} x \\ y \\ z \end{bmatrix} = \begin{bmatrix} 17 \\ -22 \\ 16 \end{bmatrix}$$ by adding 2 times the first row to the second;

$$\begin{bmatrix} 4 & 5 & 0 \\ 1 & 0 & 0 \\ 1 & 3 & 1 \end{bmatrix} \begin{bmatrix} x \\ y \\ z \end{bmatrix} = \begin{bmatrix} 17 \\ -2 \\ 16 \end{bmatrix}$$ by multiplying the second row by $1/11$;

$$\begin{bmatrix} 0 & 5 & 0 \\ 1 & 0 & 0 \\ 0 & 3 & 1 \end{bmatrix} \begin{bmatrix} x \\ y \\ z \end{bmatrix} = \begin{bmatrix} 25 \\ -2 \\ 18 \end{bmatrix}$$ by adding -4 times the second row to the first and -1 times the second row to the third;

$$\begin{bmatrix} 0 & 1 & 0 \\ 1 & 0 & 0 \\ 0 & 3 & 1 \end{bmatrix} \begin{bmatrix} x \\ y \\ z \end{bmatrix} = \begin{bmatrix} 5 \\ -2 \\ 18 \end{bmatrix}$$ by multiplying the first row by $1/5$;

$$\begin{bmatrix} 1 & 0 & 0 \\ 0 & 1 & 0 \\ 0 & 0 & 1 \end{bmatrix} \begin{bmatrix} x \\ y \\ z \end{bmatrix} = \begin{bmatrix} -2 \\ 5 \\ 3 \end{bmatrix}$$

by adding -3 times the first row to the third row and then interchanging the first two rows.

Now with matrix multiplication we obtain

$$\begin{bmatrix} x \\ y \\ z \end{bmatrix} = \begin{bmatrix} -2 \\ 5 \\ 3 \end{bmatrix} \quad \text{or} \quad \begin{matrix} x = -2 \\ y = 5 \\ z = 3. \end{matrix}$$

A second method of solving systems of linear equations makes use of the concept of the inverse of a matrix. Consider the same matrix equation

$$\begin{bmatrix} 3 & 2 & -1 \\ 6 & -1 & 3 \\ 1 & 3 & 1 \end{bmatrix} \begin{bmatrix} x \\ y \\ z \end{bmatrix} = \begin{bmatrix} 1 \\ -8 \\ 16 \end{bmatrix}.$$

We can use methods of the previous section to determine whether or not the inverse of the left matrix exists and, if so, to find that inverse and then left multiply both members of the previous matrix equation by that inverse; that is, if

$$\mathbf{A} = \begin{bmatrix} 3 & 2 & -1 \\ 6 & -1 & 3 \\ 1 & 3 & 1 \end{bmatrix}, \quad \mathbf{X} = \begin{bmatrix} x \\ y \\ z \end{bmatrix}, \quad \mathbf{K} = \begin{bmatrix} 1 \\ -8 \\ 16 \end{bmatrix},$$

then

$$\mathbf{AX} = \mathbf{K}$$
$$\mathbf{A}^{-1}\mathbf{AX} = \mathbf{A}^{-1}\mathbf{K}$$
$$\mathbf{X} = \mathbf{A}^{-1}\mathbf{K}$$

and

$$\mathbf{A}^{-1} = \begin{bmatrix} 3 & 2 & -1 \\ 6 & -1 & 3 \\ 1 & 3 & 1 \end{bmatrix}^{-1} = \frac{\operatorname{adj}\mathbf{A}}{|\mathbf{A}|} = \frac{-1}{55} \cdot \begin{bmatrix} -10 & -5 & 5 \\ -3 & 4 & -15 \\ 19 & -7 & -15 \end{bmatrix}$$

so

$$\begin{bmatrix} x \\ y \\ z \end{bmatrix} = \frac{1}{-55} \cdot \begin{bmatrix} -10 & -5 & 5 \\ -3 & 4 & -15 \\ 19 & -7 & -15 \end{bmatrix} \cdot \begin{bmatrix} 1 \\ -8 \\ 16 \end{bmatrix} = \frac{1}{-55} \cdot \begin{bmatrix} 110 \\ -275 \\ -165 \end{bmatrix} = \begin{bmatrix} -2 \\ 5 \\ 3 \end{bmatrix}.$$

Therefore, $x = -2$, $y = 5$, and $z = 3$.

Exercise Set 6.6

1. Find the reduced echelon form of each of the following matrices.

(a) $\begin{bmatrix} 5 & 1 \\ -2 & 3 \end{bmatrix}$

(b) $\begin{bmatrix} 2 & 0 \\ 5 & 3 \end{bmatrix}$

(c) $\begin{bmatrix} 0 & 1 \\ 4 & 6 \end{bmatrix}$

(d) $\begin{bmatrix} -2 & 1 \\ 3 & 2 \\ 0 & 4 \end{bmatrix}$

(e) $\begin{bmatrix} 0 & 0 \\ 4 & 1 \\ 3 & -2 \end{bmatrix}$

(f) $\begin{bmatrix} 5 & 1 & 2 \\ -2 & 4 & 3 \end{bmatrix}$

(g) $\begin{bmatrix} 3 & -1 & 3 \\ 2 & 2 & 5 \\ 4 & 8 & 9 \end{bmatrix}$

(h) $\begin{bmatrix} 4 & -3 & 1 & 7 \\ 8 & 1 & -5 & 2 \\ 0 & 1 & 3 & -2 \end{bmatrix}$

(i) $\begin{bmatrix} 3 & -7 & 4 & -1 \\ -2 & 1 & 1 & 4 \\ 3 & 4 & -1 & 9 \end{bmatrix}$

(j) $\begin{bmatrix} 5 & 1 & 3 \\ 2 & 4 & -3 \\ 9 & 9 & -3 \end{bmatrix}$

2. Use matrices to solve the following systems of equations.

(a) $x + 2y = 10$
$-2x + 3y = 1$

(b) $4x - 2y = -7$
$2x - y = -1$

(c) $x + 4y = 5$
$-2x + 3y = 1$

(d) $x + 6y = 10$
$2x + 12y = 20$

(e) $x + y = 2$
$x - y = 5$
$y - z = 3$

(f) $6x + 4y + 5z = 12$
$3x + 2y + 3z = 5$
$x + z = 3$

(g) $x + y + z = 2$
$2x + 3y + 4z = 7$
$x - 2y - 2z = -1$

(h) $5x - y + 2z = 5$
$-3x + 2y - z = 10$
$4x + y + z = 7$

(i) $x + y = -2$
$2y + z = -4$
$3x + 4z = 11$

(j) $5x - 4y - 3z = 2$
$x + y - 2z = 9$
$3x - 4z = 13$

6.8 CRAMER'S RULE

As early as the seventeenth century, determinants were used to simplify the solving of systems of equations. The work by the Japanese mathematician SekiKowa and the German philosopher-mathematician Leibnitz was not widely known or used, however, until it was re-discovered in 1750 by Professor Cramer of the University of Geneva. We shall review the method here.

Consider the system of linear equations

$$a_1 x + b_1 y = c_1$$
$$a_2 x + b_2 y = c_2.$$

If we solve this system by the method of solving simultaneous equations, we multiply both members of the first equation by b_2 and both members of the second equation by $-b_1$ and obtain

$$a_1 b_2 x + b_1 b_2 y = b_2 c_1$$
$$-a_2 b_1 x - b_1 b_2 y = -b_1 c_2$$
$$(a_1 b_2 - a_2 b_1)x = b_2 c_1 - b_1 c_2.$$

So

$$x = \frac{b_2 c_1 - b_1 c_2}{a_1 b_2 - a_2 b_1} = \frac{\begin{vmatrix} c_1 & b_1 \\ c_2 & b_2 \end{vmatrix}}{\begin{vmatrix} a_1 & b_1 \\ a_2 & b_2 \end{vmatrix}}$$

if $a_1 b_2 - a_2 b_1 \neq 0$. Thus, the denominator is the determinant of the coefficient matrix **A** of the system of equations, and the numerator is the determinant of the coefficient matrix with the coefficients of x replaced by the corresponding constants. This same method can be used not only with systems of two equations in two variables, but with systems of n equations in n variables.

Repeating the preceding process and solving for y, we obtain

$$y = \frac{\begin{vmatrix} a_1 & c_1 \\ a_2 & c_2 \end{vmatrix}}{\begin{vmatrix} a_1 & b_1 \\ a_2 & b_2 \end{vmatrix}}.$$

For the system of equations

$$a_1x + b_1y + c_1z = d_1$$
$$a_2x + b_2y + c_2z = d_2$$
$$a_3x + b_3y + c_3z = d_3$$

if the denominator is not equal to zero, the solutions are

$$x = \frac{\begin{vmatrix} d_1 & b_1 & c_1 \\ d_2 & b_2 & c_2 \\ d_3 & b_3 & c_3 \end{vmatrix}}{\begin{vmatrix} a_1 & b_1 & c_1 \\ a_2 & b_2 & c_2 \\ a_3 & b_3 & c_3 \end{vmatrix}}$$

$$y = \frac{\begin{vmatrix} a_1 & d_1 & c_1 \\ a_2 & d_2 & c_2 \\ a_3 & d_3 & c_3 \end{vmatrix}}{\begin{vmatrix} a_1 & b_1 & c_1 \\ a_2 & b_2 & c_2 \\ a_3 & b_3 & c_3 \end{vmatrix}}$$

and

$$z = \frac{\begin{vmatrix} a_1 & b_1 & d_1 \\ a_2 & b_2 & d_2 \\ a_3 & b_3 & d_3 \end{vmatrix}}{\begin{vmatrix} a_1 & b_1 & c_1 \\ a_2 & b_2 & c_2 \\ a_3 & b_3 & c_3 \end{vmatrix}}$$

Example

Using Cramer's rule, we solve the following systems of equations.

(i) $2x - y = 4$
      $3x - 4y = 1$

solution:

$$x = \frac{\begin{vmatrix} c_1 & b_1 \\ c_2 & b_2 \end{vmatrix}}{\begin{vmatrix} a_1 & b_1 \\ a_2 & b_2 \end{vmatrix}} = \frac{\begin{vmatrix} 4 & -1 \\ 1 & -4 \end{vmatrix}}{\begin{vmatrix} 2 & -1 \\ 3 & -4 \end{vmatrix}} = \frac{-15}{-5} = 3$$

and

$$y = \frac{\begin{vmatrix} a_1 & c_1 \\ a_2 & c_2 \\ a_1 & b_1 \\ a_2 & b_2 \end{vmatrix}}{} = \frac{\begin{vmatrix} 2 & 4 \\ 3 & 1 \\ 2 & -1 \\ 3 & -4 \end{vmatrix}}{} = \frac{-10}{-5} = 2.$$

(ii) $3x + 5y + 2z = \quad 2$
$4x + 2y - 3z = -1$
$2x - \ y + 5z = -11$

solution:

$$x = \frac{\begin{vmatrix} d_1 & b_1 & c_1 \\ d_2 & b_2 & c_2 \\ d_3 & b_3 & c_3 \\ a_1 & b_1 & c_1 \\ a_2 & b_2 & c_2 \\ a_3 & b_3 & c_3 \end{vmatrix}}{} = \frac{\begin{vmatrix} 2 & 5 & 2 \\ -1 & 2 & -3 \\ -11 & -1 & 5 \\ 3 & 5 & 2 \\ 4 & 2 & -3 \\ 2 & -1 & 5 \end{vmatrix}}{} = \frac{250}{-125} = -2$$

$$y = \frac{\begin{vmatrix} a_1 & d_1 & c_1 \\ a_2 & d_2 & c_2 \\ a_3 & d_3 & c_3 \\ a_1 & b_1 & c_1 \\ a_2 & b_2 & c_2 \\ a_3 & b_3 & c_3 \end{vmatrix}}{} = \frac{\begin{vmatrix} 3 & 2 & 2 \\ 4 & -1 & -3 \\ 2 & -11 & 5 \\ 3 & 5 & 2 \\ 4 & 2 & -3 \\ 2 & -1 & 5 \end{vmatrix}}{} = \frac{-250}{-125} = +2$$

$$z = \frac{\begin{vmatrix} a_1 & b_1 & d_1 \\ a_2 & b_2 & d_2 \\ a_3 & b_3 & d_3 \\ a_1 & b_1 & c_1 \\ a_2 & b_2 & c_2 \\ a_3 & b_3 & c_3 \end{vmatrix}}{} = \frac{\begin{vmatrix} 3 & 5 & 2 \\ 4 & 2 & -1 \\ 2 & -1 & -11 \\ 3 & 5 & 2 \\ 4 & 2 & -3 \\ 2 & -1 & 5 \end{vmatrix}}{} = \frac{125}{-125} = -1.$$

Exercise Set 6.7
Use Cramer's rule to solve the following systems of equations.

1. $2x + \ y = 10$
   $3x - 2y = \ 1$

2. $3x - 4y = -2$
   $x - 2y = \ 6$

3. $3x - 2y = 1$
   $4x + \ y = 5$

4. $3x + 5y = -15$
   $2x - 3y = \ \ 4$

5. $3x - 2y = 1$
   $x + 2y = 5$

6. $4x - 3y = 2$
   $2x - \ y = 2$

7. $3x - 2y = 11$
   $5x + 2y = -3$
8. $3x - 2y = 11$
   $5x + 2y = -3$
9. $3x - 4y = -2$
   $x + 2y = 3$
10. $2x + 3y = 3$
    $3x - 4y = 0$
11. $6x + y = -10$
    $3x - 2y = -10$
12. $3x - 2y = 0$
    $x + 2y = 16$
13. $2x + y + z = 1$
    $x - 2y - 3z = 1$
    $3x + 2y + 4z = 5$
14. $x + 2y - z = -3$
    $2x - y + z = 5$
    $3x + 2y - 2z = -3$

15. $2x - y + 3z = -9$
    $x + 3y - z = 10$
    $3x + y - z = 8$
16. $5x - y + 2z = 5$
    $-3x + 2y - z = 10$
    $4x + y + z = 7$
17. $x + 2y - z = -3$
    $2x + y + z = 1$
    $3x + 3y = -2$
18. $2x + z = -4$
    $3y + 4z = 11$
    $x + y = -2$
19. $x + y = 2$
    $2x - z = 1$
    $2y - 3z = -1$

6.9 SUMMARY

In this chapter, we have discussed vectors and matrices and their application to the solving of systems of equations. Vectors were treated separately, but also as a special type of matrices; that is, an n-dimensional vector is merely a $1 \times n$ matrix (a row vector) or an $n \times 1$ matrix (a column vector). The following terms were defined:

Equality (of vectors and of matrices)
Vector Addition
Scalar Multiplication (of vectors and of matrices)
Dot Product of Vectors
Matrix Addition and Subtraction
Matrix Multiplication
The Transpose of a Matrix
The Adjoint of a Matrix
The Determinant of a Matrix
The Cofactor of an Entry
Nonsingular Matrix
Elementary Row Operations
Row Equivalence
Reduced Echelon Form

Three methods were given for the solving of systems of linear equations. The first is to express the system of equations as a matrix $\mathbf{AX} = \mathbf{K}$. We then perform elementary row operations on $\mathbf{A}$ to obtain its reduced echelon form while performing the same operations in the same order on $\mathbf{K}$. The resulting product gives the solutions, if they exist, to the system of equations.

The second method of solving systems of linear equations using matrices can be used only when the system can be expressed as a matrix product $\mathbf{AX} = \mathbf{K}$, in which $\mathbf{A}$ is a nonsingular matrix. When this is the case, $\mathbf{A}^{-1}$ exists, so we left multiply both members of the equation by $\mathbf{A}^{-1}$ to obtain $\mathbf{X} = \mathbf{A}^{-1}\mathbf{K}$, which gives the unique solution to the system of equations.

The final method of solving systems of equations that was discussed is called Cramer's rule. This method allows us to solve systems of equations with determinants.

Review Exercise Set 6.8

1. If $\mathbf{u}$ and $\mathbf{v}$ are vectors defined by $\mathbf{u} = (5,-1,2)$ and $\mathbf{v} = (4,-3,2)$, find the following.

    (a) $\mathbf{u} + \mathbf{v}$ (f) $|\mathbf{u}|$
    (b) $\mathbf{u} - \mathbf{v}$ (g) $|\mathbf{v}|$
    (c) $2\mathbf{u} + 3\mathbf{v}$ (h) $|\mathbf{u} + \mathbf{v}|$
    (d) $\mathbf{u} \cdot \mathbf{v}$ (i) $|\mathbf{u}| + |\mathbf{v}|$
    (e) $\mathbf{v} \cdot \mathbf{u}$ (j) $(\mathbf{u} + \mathbf{v}) \cdot (\mathbf{u} - \mathbf{v})$

2. If

$$\mathbf{A} = \begin{bmatrix} 5 & -1 & 2 \\ 3 & 4 & 2 \\ 4 & 0 & 1 \end{bmatrix} \quad \text{and} \quad \mathbf{B} = \begin{bmatrix} 1 & -3 & 2 \\ 7 & 5 & -4 \\ 6 & -1 & 0 \end{bmatrix}$$

find the following.

    (a) $\mathbf{A} + \mathbf{B}$ (h) $|\mathbf{A}|$
    (b) $3\mathbf{A}$ (i) $\mathbf{A}^{-1}$
    (c) $2\mathbf{B} - \mathbf{A}$ (j) $\mathbf{B}^{-1}\mathbf{A}^{-1}$
    (d) $\mathbf{AB}$ (k) $(\mathbf{AB})^{-1}$
    (e) $\mathbf{BA}$ (l) $(\mathbf{AB})^{t}$
    (f) $\mathbf{A}^{t}$ (m) $\mathbf{A}^{t}\mathbf{B}^{t}$
    (g) $\text{adj}\mathbf{A}$ (n) $\mathbf{B}^{t}\mathbf{A}^{t}$
                                               (o) $\mathbf{A}^{-1}\mathbf{B}^{-1}$

3. Solve the following systems of equations using methods discussed in this chapter.

(a) $4x + y = 3$
    $8x - y = -3$

(b) $2x - y = 4$
    $x - 3y = -3$

(c) $4x + y = 3$
    $8x - y = -3$

(d) $2x - 4y = 7$
    $x - 2y = 1$

(e) $4x + 5y = -1$
    $6x + 2y = 4$

(f) $3x - 2y = -2$
    $9x - 6y = -6$

(g) $3x - 2y - z = 1$
    $2x + 3y - z = 4$
    $x - y + 2z = 7$

(h) $3x - 2y + 4z = 1$
    $4x + y - 5z = 2$
    $2x - 3y + z = -6$

(i) $2x + 2y + z = 1$
    $x - y + 6z = 1$
    $3x + 2y - z = -4$

(j) $3x + 2y - z = 9$
    $-4x + 5y + 3z = 10$
    $-2x + 3y + 2z = 8$

(k) $2x - 6y + 3z = -12$
    $3x - 2y + 5z = -4$
    $4x + 5y - 2z = 10$

(l) $3x + y - 4z = -6$
    $2x + 3y + z = 5$
    $5x + 4y - 3z = -2$

chapter

7

LINEAR PROGRAMMING AND THE THEORY OF GAMES

7.1 INTRODUCTION

Not many years ago, mathematics was considered by many to be largely just a tool for the physical sciences. Later, with the emphasis on mathematics and science during the space race of the 1950s, mathematics finally became recognized by nonmathematicians as a science in, and of, itself — very pure in nature — with applied mathematics as only one small area of the overall discipline. Since that time, the emphasis on applied mathematics has increased as more and more mathematical tools have been found to be not only useful but even vital to social and biological sciences as well as to business and economics. Economic problems of efficiency come up constantly, in which we must meet desired goals with limited resources. Often, the primary objective is to plan a course of action that will accomplish a task most efficiently with the least possible labor force or monetary investment, within the restrictions of limited ingredients and facilities. Such problems can be very complicated and difficult to solve, but when the information can be expressed by linear equations or linear inequalities, then we can solve such problems by methods referred to as *linear programming*.

Some of the methods in this chapter assume an ability to solve and graph inequalities. If a review of such methods is found to be necessary, such a review can be found in Appendix A.

7.2 GEOMETRIC SOLUTIONS

One type of linear programming problem, which is common to many businesses and industries, consists of "mixing" available produce in such a way as to maximize profit. For example, consider a vegetable canning company with 599 pounds of peas and 499 pounds of diced carrots on hand. They can sell mixed vegetables in two ways, both in 3-pound cans. Mixture A has 2 pounds of peas and 1 pound of carrots to every can, and mixture B has 1.5 pounds of peas and 1.5 pounds of carrots to every can. Mixture A can be sold at $1.50 per can, and mixture B can be sold at $1.20 per can. How should the vegetables be canned and sold for maximum profit? At first, it may seem that they should all be canned according to mixture A, since it sells at a much higher price. This is not true, however, since this method of canning would not use all of the available resources; that is, 200 pounds of carrots would be wasted.

We let x represent the number of cans of mixture A and y represent the number of cans of mixture B. We now have the following constraints placed on the problem.

$x \geq 0$ (We cannot have less than 0 can of either mixture.)
$y \geq 0$
$2x + 1.5y \leq 599$ (The sum of the pounds of peas in the two mixtures cannot exceed the pounds of peas available.)
$x + 1.5y \leq 499$ (The sum of the pounds of carrots in the two mixtures cannot exceed the pounds of carrots available.)

The graph of all possible solutions is the following shaded area, which satisfies all of the constraints. (See Figure 7.1.) The income from the total sales is c in the equation

$$\$1.50x + \$1.20y = c.$$

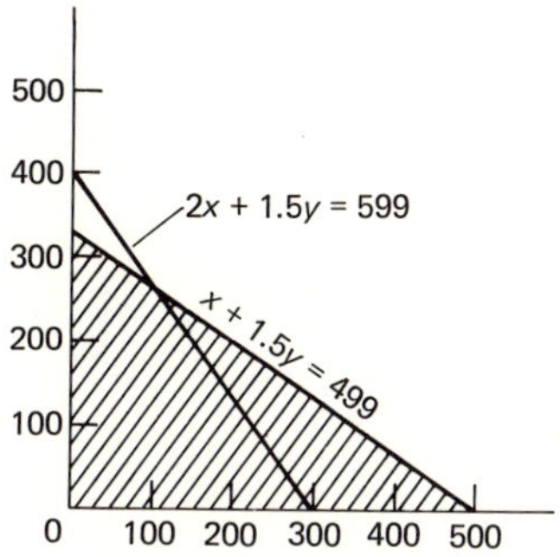

FIGURE 7.1

Since the ratio of x to y is constant, all lines satisfying this equation are parallel. All points on such lines that are in the shaded region are solutions to the equation. Also, the greater the distance between the line and the origin, the greater the income. For example, the lines $1.50x + 1.20y = 300$, $1.50x + 1.20y = 150$, and $1.50x + 1.20y = 599$ are all equations of the form $1.50x + 1.20y = c$. As Figure 7.2 illustrates, the lines illustrating incomes of $150 and $300 have solutions, with the line illustrating the $300 income lying at a greater distance from the origin, but the line illustrating an income of $600 does not intersect the shaded region, so it has no feasible solution.

Maximum profits come at the point or points where a solution line can intersect the area of feasible solutions at the greatest distance possible from the origin. In the case of this problem, that point, as illustrated in Figure 7.3, is the point where the lines $x + 1.5y = 499$ and $2x + 1.5y = 599$ intersect. Solving the system of equations, we obtain $x = 100$ and $y = 266$. Therefore, by selling 100 cans of mixture A and 266 cans of mixture B, we get a maximum income of

$$\$1.50(100) + \$1.20(266) = \$469.20.$$

The equation of the line containing the maximum point is $1.50x + 1.20y = 469.20$.

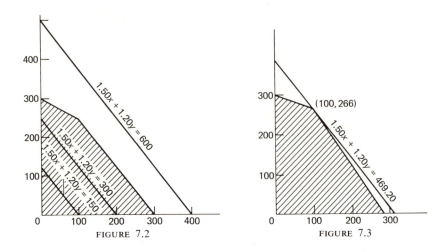

FIGURE 7.2   FIGURE 7.3

In such problems, only at vertices, or points of intersection of the boundaries of the constraints, will all available resources be used. Thus, *any maximum or minimum solution to such a linear programming problem will always include a vertex of the polygonal region determined by the constraints.* If it includes more than one vertex, then it also includes the entire segment between them.

Example

A store manager wishes to mix Christmas nuts that he has on hand in a way that will bring him the greatest possible income. All he has on hand are 600 pounds of walnuts, 600 pounds of Brazil nuts, and 500 pounds of almonds. He knows two ways in which he can mix the nuts to sell. Method A mixes 2 pounds of walnuts and 2 pounds of almonds to each pound of Brazil nuts; that is, the mixture is $\frac{2}{5}$ walnuts, $\frac{2}{5}$ almonds, and $\frac{1}{5}$ Brazil nuts. Method B mixes 2 pounds of walnuts and 1 pound of almonds to 3 pounds of Brazil nuts; that is, the mixture is $\frac{1}{3}$ walnuts, $\frac{1}{6}$ almonds, and $\frac{1}{2}$ Brazil nuts. He can sell mixture A for $0.60 per pound and mixture B for $1.00 per pound and sell as many pounds of each mixture as he has available.

If we let x be the pounds of mixture A and y be the pounds of mixture B, the constraints are the following:

$$x \geqslant 0$$
$$y \geqslant 0$$
$$\tfrac{2}{5}x + \tfrac{1}{3}y \leqslant 600$$
$$\tfrac{2}{5}x + \tfrac{1}{6}y \leqslant 500$$

and

$$\tfrac{1}{5}x + \tfrac{1}{2}y \leqslant 600.$$

The last three constraints can also be written $6x + 5y \leqslant 9000$, $12x + 5y \leqslant 15,000$, and $2x + 5y \leqslant 6000$. The graph of feasible solutions is as shown in Figure 7.4.

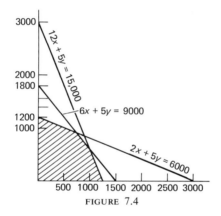

FIGURE 7.4

The desired solution is the largest possible value of c such that the equation

$$0.60x + 1.00y = c$$

intersects the shaded area of the graph as shown in Figure 7.5. Since the ratio of x to y is constant, the line containing the desired solution is parallel to

$$0.60x + 1.00y = 600.$$

Thus, the values for x and y at the point of maximum income are the coordinates of the point of intersection of $6x + 5y = 9000$

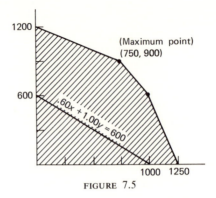

FIGURE 7.5

and $2x + 5y = 6000$, that is, $(750, 900)$. So the maximum income is

$$\$0.60(750) + \$1.00(900) = \$1350.00$$

and is obtained by selling 750 pounds of mixture A and 900 pounds of mixture B.

Example

A dietitian must prepare two foods in such a way as to obtain a minimum of 1000 grams of protein and 1500 grams of carbohydrates. Each unit of food A contains 15 grams of protein and 25 grams of carbohydrate and costs 20¢. Each unit of food B contains 20 grams of protein and 20 grams of carbohydrate and costs 25¢. He wishes to obtain the least expensive mixture possible that still satisfies the constraints.

If x represents the number of units of food A in the mixture and y represents the number of units of food B in the mixture, then the constraints are the following:

$$x \geqslant 0$$
$$y \geqslant 0$$
$$15x + 20y \geqslant 1000 \quad \text{or more simply } 3x + 4y \geqslant 200$$
$$25x + 20y \geqslant 1500 \quad \text{or more simply } 5x + 4y \geqslant 300.$$

The constraints are illustrated in Figure 7.6, in which the shaded area represents the feasible solutions. The cost of the mixture is

$$c = \$0.20x + \$0.25y$$

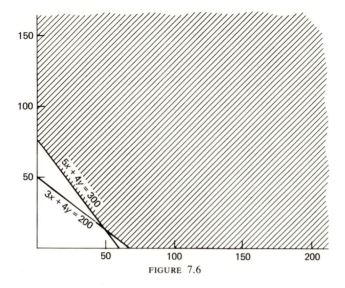

FIGURE 7.6

so the minimum cost is the point or points *closest* to the origin at which a line $c = 0.20x + 0.25y$ intersects the area of feasible solutions. The solution line is parallel to $20x + 25y = 1000$, which is illustrated in Figure 7.7. Thus, the minimum solution is at the vertex $(50, 12\frac{1}{2})$, which is the point of intersection of the lines $3x + 4y = 200$ and $5x + 4y = 300$. Thus, the least expensive mixture possible that still satisfies the given constraints contains 50 units of food A and $12\frac{1}{2}$ units of food B and costs a total of

$$\$0.20(50) + \$0.25(12\tfrac{1}{2}) = \$10.00 + \$3.12\tfrac{1}{2} \approx \$13.13.$$

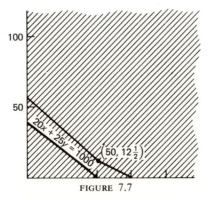

FIGURE 7.7

The previous examples have used graphical methods of solving problems, which could also have been solved without graphs. Since maximum and minimum values are always at vertices, all that is necessary to determine these values is to find the coordinates of each vertex and compute the value of the function at that point. For example, in the problem of mixing vegetables, the constraints were

$$x \geq 0$$
$$y \geq 0$$
$$2x + 1.5y \leq 599$$
$$x + 1.5y \leq 499.$$

Thus, the vertices of the area of feasible solutions are the points of intersection of

$$x = 0 \quad \text{and} \quad y = 0, \text{ that is, } (0,0);$$
$$y = 0 \quad \text{and} \quad 2x + 1.5y = 599, \text{ that is, } (299\tfrac{1}{2},0);$$
$$x = 0 \quad \text{and} \quad x + 1.5y = 499, \text{ that is, } (0,332\tfrac{2}{3});$$

and

$$x + 1.5y = 499 \quad \text{and} \quad 2x + 1.5y = 599, \text{ that is, } (100,266).$$

By evaluating the equation $c = \$1.50x + \$1.20y$ at each of these points, we obtain the following table in which we see that $(0,0)$ is the minimum point and $(100,266)$ is the maximum point. The other example can also be solved in this manner.

Vertex	$1.50x + 1.20y$
$(0,0)$	
$(299\tfrac{1}{2},0)$	449.25
$(0,332\tfrac{2}{3})$	399.20
$(100,266)$	469.20

Linear programming problems involving three variables are difficult to graph, and problems involving more than three variables must be solved by other means. Thus, algebraically determining the vertices and evaluating at those vertices is about the only means we have at this point of solving such programs. Even this, however, is a cumbersome method of solving such programs, since they usually have a large number of vertices. For example, the three-dimensional figure illustrated by Figure 7.8 has 10 vertices, the determination of which would require the solving of 10 systems of 3 equations each in 3 variables. Fortunately, other methods for solving such programs have been developed. One such method will be discussed in the next section of this chapter.

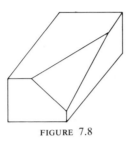

FIGURE 7.8

Exercise Set 7.1

1. Graph the following constraints and find the maximum and minimum values for c and the coordinates (x,y) at such values, where $c = 50x + 40y$.

    (a) $x \geqslant 0$
     $y \geqslant 0$
     $x + y \leqslant 30$
     $x + 10y \leqslant 100$

    (b) $x \geqslant 0$
     $y \geqslant 0$
     $5x + y \leqslant 40$
     $x + 5y \leqslant 40$

    (c) $x \geqslant 0$
     $y \geqslant 0$
     $x \leqslant 30$
     $y \leqslant 15$

    (d) $x \geqslant 0$
     $y \geqslant 1$
     $x + y \geqslant 5$
     $x \leqslant 8$
     $y \leqslant 5$

    (e) $x \geqslant 0$
     $y \geqslant 0$

     $x \leqslant 5$
     $y \leqslant 5$
     $x + y \leqslant 8$

    (f) $x \geqslant 0$
     $y \geqslant 0$
     $2x + y \geqslant 5$
     $x - y \geqslant 3$

    (g) $x \geqslant 0$
     $y \geqslant 0$
     $x + y \geqslant 5$
     $x \leqslant 8$
     $2x + 5y \leqslant 25$

    (h) $x \geqslant 0$
     $y \geqslant 2$
     $x \leqslant 10$
     $y \leqslant 12$
     $2x + y \geqslant 6$
     $2x + y \leqslant 12$

(i) $x \geqslant 3$ (j) $x \geqslant 1$
     $y \geqslant 2$ $y \geqslant 3$
     $x \leqslant 7$ $y \leqslant 8$
     $y \leqslant 10$ $x + y \geqslant 6$
     $5x + 4y \geqslant 5$ $x + y \leqslant 15$
                                               $y \leqslant 2x + 3$
                                               $y \geqslant 2x - 15$

2. A canning company cans and sells both vegetable beef soup and beef stew. The only difference in their soup and their stew is that the soup is 1 part beef and 4 parts vegetables, while the stew is 2 parts beef and 3 parts vegetables. The soup sells at 30¢ per pound, while the stew sells at 50¢ per pound. If the company has 7000 pounds of vegetables and 3000 pounds of beef available, how can the resources available best be canned to obtain maximum income?

3. A producer of breakfast foods makes two types of cereals. The first has 2 ounces of oats, 3 ounces of barley, and 5 ounces of wheat per package, while the second has 5 ounces of oats, 3 ounces of barley, and 2 ounces of wheat per package. He has 44 pounds of wheat, 30 pounds of barley, and 41 pounds of oats on hand. Find the number of packages of each cereal that he should produce to maximize his income if the costs are the following. Also find the maximum income in each case.

    (a) The first type of cereal sells for 60¢ per package, and the second sells for 40¢ per package.
    (b) The first cereal sells for 40¢ per package, and the second sells for 80¢ per package.
    (c) The first cereal sells for $1.00 per package, and the second sells for 25¢ per package.
    (d) The first cereal sells for 25¢ per package, and the second sells for $1.00 per package.

4. A contractor has 200 units of glass, 300 units of lumber, and 400 units of concrete available. He also has all of the rest of the material needed for completing homes, but there is no difference in the cost of the other supplies for the different types of homes he builds. He builds only two types of homes. The first requires 2 units of glass, 5 units of lumber, and 8 units of concrete. The second requires 5 units of glass, 6 units of lumber, and 5 units of concrete. Find the number of homes of each type that he should

build to maximize his income when they sell for the following prices. Find the maximum income in each case.

(a) The first type home sells for $15,000.00 and the second for $20,000.00.

(b) The first type home sells for $20,000.00 and the second for $40,000.00.

(c) The first type home sells for $25,000.00 and the second for $25,000.00.

(d) The first type home sells for $25,000.00 and the second for $30,000.00.

(e) The first type home sells for $20,000.00 and the second for $15,000.00.

5. A sewing factory has the capacity to produce 480 sweaters per day. Two kinds of sweaters are produced. Type A requires 12 ounces of wool, and type B requires 18 ounces of wool. On a particular day, there are only 450 pounds of wool available. Find the number of sweaters of each type that should be produced on that day to maximize profit when the profit per sweater is

(a) $6.00 for type A and $6.00 for type B;

(b) $12.00 for type A and $9.00 for type B;

(c) $4.00 for type A and $8.00 for type B.

6. An alfalfa farmer has 200 acres of alfalfa and a problem of dotter spreading through his fields. He knows that he can control the dotter problem by cutting 3 crops for hay and leaving none of the alfalfa for seed, but this will also cut his profits considerably for that year. Thus, he decides that he will cut 3 crops of hay on as much of the land as he possibly can and raise only enough alfalfa seed to net $24,000.00 for his living and machinery expenses. On that portion of the farm from which he cuts 3 crops of hay, he receives a net profit of $80.00 per acre. On the portion from which he raises seed, he receives a net profit of $240.00 per acre. How many acres of hay and how many acres of seed should he produce to minimize the number of acres of seed without dropping profits below $24,000.00?

7. Once the farmer in Problem 6 has solved his dotter problem, he plans to enter the cattle business, so will no longer be selling all of his hay. In fact, he finds that to feed the number of cattle he plans to buy, he will need at least 200 tons of hay per year, but will sell all of the hay that he raises above that amount at $30.00

per ton. He raises 6 tons per acre on the land from which he cuts 3 crops and 3 tons per acre on the land from which he raises seed (he cuts the first crop of hay and leaves the rest for seed). He still nets $240.00 per acre on the seed that he raises. If he wishes to maximize profits, how many acres of seed should he raise? What is his maximum profit?

7.3 THE SIMPLEX METHOD

Up to this point, we have discussed two methods of solving linear programs — graphically and through evaluating the function at the vertices. When several variables are involved, however, a program cannot be solved graphically, and when there are many vertices, the method of evaluation of the vertices becomes tedious and impractical. For example, if a program involves 6 variables and 4 constraints in addition to the constraints that all of the variables be nonnegative, then there are 210 vertices to be evaluated, each of which would be a solution to a system of 6 equations in 6 variables. As the number of constraints increases, so does the number of vertices.

Thus, we see that the methods already discussed for solving linear programs are inadequate for the solving of more complex linear programs. Because of the inadequacy of these methods, we introduce an alternative method at this point. This method, called the *simplex method,* or the *simplex algorithm,* is rapidly gaining popularity because of its adaptability to computers and the ever-increasing use of computers in today's business world.

The simplex method appeared in the United States in 1947 as a result of research by George B. Dantzig and his associates in the U.S. Department of the Air Force and soon became recognized as the most effective method for solving linear programming problems.

In general, the simplex method of solving a linear program is a highly efficient trial-and-error method. It begins with a *feasible solution,* that is, any solution that satisfies all of the constraints, and determines whether or not that solution is an *optimal solution,* that is, the most desirable solution, which will be a maximum or minimum value of the function in question. If the feasible solution is not optimal, then the simplex method indicates the direction that must be taken to arrive at the solution that is optimal — if such a solution exists. If an optimal solution does not exist, this will be indicated early in the problem, and if it does exist, it will eventually be reached.

Consider a linear program in 3 variables x_1, x_2, and x_3 with the constraints $x_1 \geq 0$, $x_2 \geq 0$, $x_3 \geq 0$ and

$$x_1 + 4x_2 - 2x_3 \leq 10$$
$$3x_2 + x_3 \leq 7$$
$$2x_1 - 3x_2 + x_3 \leq 12.$$

The problem is to maximize the value v, where $v = 2x_1 + 3x_2 + x_3$.

The constraints of the program are inequalities, but can also be expressed as equations. For example, according to the definition of $\leq$, the inequality

$$x_1 + 4x_2 - 2x_3 \leq 10$$

means that there is a number $x_4 \geq 0$ such that

$$x_1 + 4x_2 - 2x_3 + x_4 = 10.$$

Likewise, there exist numbers $x_5 \geq 0$ and $x_6 \geq 0$ such that

$$3x_2 + x_3 + x_5 = 7$$

and

$$2x_1 - 3x_2 + x_3 + x_6 = 12.$$

We call the variables x_4, x_5, and x_6 introduced in this way *slack variables*, since they take up the "slack" between the left and right numbers of the inequalities. We now have the following system of equations:

$$x_1 + 4x_2 - 2x_3 + x_4 = 10$$
$$3x_2 + x_3 + x_5 = 7$$
$$2x_1 - 3x_2 + x_3 + x_6 = 12$$

which can be expressed as the matrix equation

$$\begin{bmatrix} 1 & 4 & -2 & 1 & 0 & 0 \\ 0 & 3 & 1 & 0 & 1 & 0 \\ 2 & -3 & 1 & 0 & 0 & 1 \end{bmatrix} \cdot \begin{bmatrix} x_1 \\ x_2 \\ x_3 \\ x_4 \\ x_5 \\ x_6 \end{bmatrix} = \begin{bmatrix} 10 \\ 7 \\ 12 \end{bmatrix}$$

which is often written in the more abbreviated form

$$\begin{array}{cccccc} x_1 & x_2 & x_3 & x_4 & x_5 & x_6 \\ \end{array}$$
$$\left[\begin{array}{cccccc|c} 1 & 4 & -2 & 1 & 0 & 0 & 10 \\ 0 & 3 & 1 & 0 & 1 & 0 & 7 \\ 2 & -3 & 1 & 0 & 0 & 1 & 12 \end{array}\right].$$

We can now use elementary row operations to solve this system for the variables x_1, x_2, and x_3 in terms of the slack variables x_4, x_5, and x_6. To do so, we must obtain a 1 in each of the circled positions of the matrix with the remainder

$$\begin{array}{cccccc} x_1 & x_2 & x_3 & x_4 & x_5 & x_6 \\ \end{array}$$
$$\left[\begin{array}{cccccc|c} ① & 4 & -2 & 1 & 0 & 0 & 10 \\ 0 & ③ & 1 & 0 & 1 & 0 & 7 \\ 2 & -3 & ① & 0 & 0 & 1 & 12 \end{array}\right]$$

of the entries in those same columns being zeros. This is called a *complete elimination process*. First, we choose one of the circled entries as the *pivot entry*. The row containing that entry is called the *pivot row*, and the column containing it is called the *pivot column*. The pivot entry is commonly circled. Now we choose to use the 1 in the first row and the first column as the first pivot entry. By adding -2 times the first row to the third, we obtain the matrix

$$\begin{array}{cccccc} x_1 & x_2 & x_3 & x_4 & x_5 & x_6 \\ \end{array}$$
$$\left[\begin{array}{cccccc|c} ① & 4 & -2 & 1 & 0 & 0 & 10 \\ 0 & 3 & 1 & 0 & 1 & 0 & 7 \\ 0 & -11 & 5 & -2 & 0 & 1 & -8 \end{array}\right].$$

By continuing the elimination process, using 3 as the pivot entry, we first multiply the second row by $\frac{1}{3}$, and then add -4 times the new second row to the first row and 11 times the new second row to the third row to obtain the matrix

$$
\begin{array}{cccccc}
x_1 & x_2 & x_3 & x_4 & x_5 & x_6
\end{array}
$$

$$
\begin{bmatrix}
1 & 0 & \dfrac{-10}{3} & 1 & \dfrac{-4}{3} & 0 & \bigg| & \dfrac{2}{3} \\[3mm]
0 & 1 & \dfrac{1}{3} & 0 & \dfrac{1}{3} & 0 & \bigg| & \dfrac{7}{3} \\[3mm]
0 & 0 & \dfrac{26}{3} & -2 & \dfrac{11}{3} & 1 & \bigg| & \dfrac{53}{3}
\end{bmatrix}.
$$

We can now complete the elimination process using $26/3$ as the pivot entry by first multiplying the third row by $3/26$ and then adding $-1/3$ times the new third row to the second row and $10/3$ times the new third row to the first row to obtain the matrix

$$
\begin{array}{cccccc}
x_1 & x_2 & x_3 & x_4 & x_5 & x_6
\end{array}
$$

$$
\begin{bmatrix}
1 & 0 & 0 & \dfrac{3}{13} & \dfrac{1}{13} & \dfrac{5}{13} & \bigg| & \dfrac{97}{13} \\[3mm]
0 & 1 & 0 & \dfrac{1}{13} & \dfrac{5}{26} & \dfrac{-1}{26} & \bigg| & \dfrac{43}{26} \\[3mm]
0 & 0 & 1 & \dfrac{-3}{13} & \dfrac{11}{26} & \dfrac{3}{26} & \bigg| & \dfrac{53}{26}
\end{bmatrix}.
$$

Thus, we have the system of equations

$$
x_1 = \frac{97}{13} - \frac{3}{13}x_4 - \frac{1}{13}x_5 - \frac{5}{13}x_6
$$

$$
x_2 = \frac{43}{26} - \frac{1}{13}x_4 - \frac{5}{26}x_5 - \frac{1}{26}x_6
$$

$$
x_3 = \frac{53}{26} + \frac{3}{13}x_4 - \frac{11}{26}x_5 - \frac{3}{26}x_6
$$

which is equivalent to our original system. This system, however, has an infinite number of solutions, which can be obtained by assigning values to x_4, x_5, and x_6. We wish, however, to maximize v, where

$$v = 2x_1 + 3x_2 + x_3$$

$$= 2\left(\frac{97}{13} - \frac{3}{13}x_4 - \frac{1}{13}x_5 - \frac{5}{13}x_6\right)$$

$$+ 3\left(\frac{43}{26} - \frac{1}{13}x_4 - \frac{5}{26}x_5 - \frac{1}{26}x_6\right) + \left(\frac{53}{26} + \frac{3}{13}x_4 - \frac{11}{26}x_5 - \frac{3}{26}x_6\right)$$

$$= \frac{285}{13} - \frac{6}{13}x_4 - \frac{28}{13}x_5 - \frac{16}{13}x_6.$$

At this point, it is obvious that the maximum value for v is $285/13$ and occurs when $x_4 = 0$, $x_5 = 0$, and $x_6 = 0$. Therefore, the values of x_1, x_2, and x_3 at the maximum point are

$$x_1 = \frac{93}{13}$$

$$x_2 = \frac{43}{26}$$

$$x_3 = \frac{53}{26}.$$

These same results may be found by a somewhat more abbreviated and systematic method by including the function that we wish to maximize in the form

$$-2x_1 - 3x_2 - x_3 + v = 0$$

as the last row of the original matrix equation in the following way. This matrix is called the *simplex tableau*.

$$\begin{array}{ccccccc} x_1 & x_2 & x_3 & x_4 & x_5 & x_6 & v \\ \begin{bmatrix} 1 & 4 & -3 & 1 & 0 & 0 & 0 \\ 0 & 3 & 1 & 0 & 1 & 0 & 0 \\ 2 & -3 & 1 & 0 & 0 & 1 & 0 \\ -2 & -3 & -1 & 0 & 0 & 0 & 1 \end{bmatrix} & \begin{matrix} 10 \\ 7 \\ 12 \\ 0 \end{matrix} \end{array}.$$

The program can now be solved by the following steps.

1) Locate the negative number with the greatest absolute value in the last row on the left side of the vertical bar.

2) Divide each positive entry in that column into the extreme right entry of the same row.
3) Select the entry that is the divisor of the *smallest* quotient as the pivot entry.
4) Complete the elimination process for that pivot entry.
5) Repeat the process until there are no negative entries remaining in the last row to the left of the vertical bar.

Example

The nut-mixing example of the previous section has the constraints

$$12x_1 + 5x_2 \leq 15{,}000$$
$$6x_1 + 5x_2 \leq 9000$$
$$2x_1 + 5x_2 \leq 6000$$

besides the nonnegative constraints $x_1 \geq 0$, $x_2 \geq 0$, and $x_3 \geq 0$. The function that we wish to maximize is

$$60x_1 + 100x_2 = C.$$

To solve this program by the simplex method, we first introduce the slack variables $x_3 \geq 0$, $x_4 \geq 0$, and $x_5 \geq 0$ such that

$$12x_1 + 5x_2 + x_3 = 15{,}000$$
$$6x_1 + 5x_2 + x_4 = 9000$$
$$2x_1 + 5x_2 + x_5 = 6000.$$

We now express the system of equations in the simplex tableau

x_1	x_2	x_3	x_4	x_5	C	
12	5	1	0	0	0	15,000
6	5	0	1	0	0	9000
2	5	0	0	1	0	6000
-60	-100	0	0	0	1	0

1) The negative entry in the last row with the greatest absolute value is -100.

2) Dividing each positive entry into the extreme right entry of the same row, we obtain 3000 in the first row, 1800 in the second row, and 1200 in the third row.

3) Since 1200 is the least of the three quotients, we select the entry in the second column and the third row as the pivot entry.

4) We now complete the elimination process for the chosen pivot in the matrix

x_1	x_2	x_3	x_4	x_5	C	
12	5	1	0	0	0	15,000
6	5	0	1	0	0	9000
2	⑤	0	0	1	0	6000
-60	-100	0	0	0	1	0

to obtain the matrix

x_1	x_2	x_3	x_4	x_5	C	
10	0	1	0	-1	0	9000
4	0	0	1	-1	0	3000
$\frac{2}{5}$	1	0	0	$\frac{1}{5}$	0	1200
-20	0	0	0	20	1	120,000

Repeating the process, we find that the only negative entry remaining in the last row is -20 in the left column. Dividing each entry in the right column by the corresponding positive entry in the left column, we obtain 900 in the first row, 750 in the second row, and 3000 in the third row. Thus, we select the entry in the first column and the second row as the pivot entry. We now complete the elimination process for the chosen pivot entry the matrix

x_1	x_2	x_3	x_4	x_5	C	
10	0	1	0	-1	0	9000
④	0	0	1	-1	0	3000
$\frac{2}{5}$	1	0	0	$\frac{1}{5}$	0	1200
-20	0	0	0	20	1	120,000

to obtain the matrix

$$\begin{array}{ccccccc}
x_1 & x_2 & x_3 & x_4 & x_5 & C & \\
\end{array}$$
$$\begin{bmatrix}
0 & 0 & 1 & -\frac{5}{2} & \frac{3}{2} & 0 & 1500 \\
1 & 0 & 0 & \frac{1}{4} & -\frac{1}{4} & 0 & 750 \\
0 & 1 & 0 & -\frac{1}{10} & \frac{3}{10} & 0 & 900 \\
0 & 0 & 0 & 5 & 15 & 1 & 135{,}000
\end{bmatrix}$$

Since there are no more negative numbers in the last row, we conclude that the system cannot be improved. From the matrix, we now obtain the equation

$$5x_4 + 15x_5 + C = 135{,}000$$

or

$$C = 135{,}000 - 5x_4 - 15x_5.$$

Since we want C to have its maximum possible value and x_4 and x_5 are nonnegative, we let $x_4 = 0$ and $x_5 = 0$ to obtain the maximum value $C = 135{,}000$. Therefore, we have

$$x_1 = 750 - \tfrac{1}{4}x_4 + \tfrac{1}{4}x_5 = 750$$

and

$$x_2 = 900 + \tfrac{1}{10}x_4 - \tfrac{3}{10}x_5 = 900$$

at the maximum value of C.

To find a minimum value of some function f, we need to find the maximum value of $-f$. Often this involves no new procedure whatever, but this is not always the case. When $f = ax_1 + bx_2 + Cx_3$ and $a, b,$ and c are all positive, $-f = -ax_1 - bx_2 - cx_3$, so $ax_1 + bx_2 + cx_3 + (-f) = 0$, and none of the entries of the last row is negative — thus indicating no feasible solution with which to begin the simplex method. In such cases, additional steps are needed to solve the programs.

Example

A dietitian must mix three foods such that the mixture will contain at least 50 ounces of protein, 60 ounces of carbohydrate, and 65 ounces of fat. Each unit of food A contains 10 ounces of protein, 4 ounces of carbohydrate, and 6 ounces of fat and costs 80¢. Each unit of food B contains 2 ounces of protein, 3 ounces of carbohydrate, and 4 ounces of fat and costs 20¢. Each unit of food C contains 1 ounce of protein, 6 ounces of carbohydrate, and 4 ounces of fat and costs 15¢. How many units of each food should be mixed to obtain the desired ingredients at the least possible cost?

solution:

If we let x_1 represent the number of units of food A, x_2 represent the number of units of food B, and x_3 represent the number of units of food C in the mixture, we have the following constraints

$$x_1 \geq 0$$
$$x_2 \geq 0$$
$$x_3 \geq 0$$
$$10x_1 + 2x_2 + x_3 \geq 50$$
$$4x_1 + 3x_2 + 6x_3 \geq 60$$
$$6x_1 + 4x_2 + 4x_3 \geq 65.$$

The function that we wish to *minimize* is

$$C = 80x_1 + 20x_2 + 15x_3.$$

Therefore, the function we wish to *maximize* is

$$m = -C = -80x_1 - 20x_2 - 15x_3.$$

To solve the program, we introduce the slack variable $x_4 \geq 0$, $x_5 \geq 0$, and $x_6 \geq 0$ such that

$$
\begin{aligned}
10x_1 + 2x_2 + x_3 - x_4 &= 50 \\
4x_1 + 3x_2 + 6x_3 \quad - x_5 &= 60 \\
6x_1 + 4x_2 + 4x_3 \quad\quad - x_6 &= 65 \\
80x_1 + 20x_2 + 15x_3 \quad\quad\quad + m &= 0.
\end{aligned}
$$

The simplex tableau becomes

$$
\begin{array}{ccccccc}
x_1 & x_2 & x_3 & x_4 & x_5 & x_6 & m \\
\end{array}
$$

$$
\left[
\begin{array}{ccccccc|c}
10 & 2 & 1 & -1 & 0 & 0 & 0 & 50 \\
4 & 3 & 6 & 0 & -1 & 0 & 0 & 60 \\
6 & 4 & 4 & 0 & 0 & -1 & 0 & 65 \\
\hline
80 & 20 & 15 & 0 & 0 & 0 & 1 & 0
\end{array}
\right].
$$

Note that none of the entries of the last row is negative, which may leave the impression that there is no maximum value for m, yet in considering the problem itself, we find that this conclusion is not feasible. This dilemma often occurs with problems such as this in which constraints are expressed in terms of $\geqslant$, and the problem is ultimately a minimum problem rather than a maximum problem. In such cases, we find it necessary to introduce another variable, called an *artificial variable*. An artificial variable differs from a slack variable in that an artificial variable always disappears in the process of solving the program, while a slack variable may actually take on a value in the solution of the program.

We introduce three new "artificial variables" a, b, and c into the program (one for each row except the last row). Thus, we have

$$
\begin{array}{cccccccccc}
x_1 & x_2 & x_3 & x_4 & x_5 & x_6 & a & b & c & m \\
\end{array}
$$

$$
\left[
\begin{array}{cccccccccc|c}
10 & 2 & 1 & -1 & 0 & 0 & 1 & 0 & 0 & 0 & 50 \\
4 & 3 & 6 & 0 & -1 & 0 & 0 & 1 & 0 & 0 & 60 \\
6 & 4 & 4 & 0 & 0 & -1 & 0 & 0 & 1 & 0 & 65 \\
\hline
80 & 20 & 15 & 0 & 0 & 0 & 0 & 0 & 0 & 1 & 0
\end{array}
\right].
$$

This may seem to change the nature of the program drastically, but we avoid this by subtracting from m the quantity $pa + pb + pc$, where p is an unspecified but "large" positive number. We now replace m by the artificial variable m', where

$$
\begin{aligned}
m' &= m - pa - pb - pc \\
&= -80x_1 - 20x_2 - 15x_3 - pa - pb - pc.
\end{aligned}
$$

We enter this into the simplex tableau with the last equation listed as a double row, with that part involving p written below that which does not involve p.

x_1	x_2	x_3	x_4	x_5	x_6	a	b	c	m'	
10	2	1	-1	0	0	1	0	0	0	50
4	3	6	0	-1	0	0	1	0	0	60
6	4	4	0	0	-1	0	0	1	0	65
80	20	15	0	0	0	0	0	0	1	0
0	0	0	0	0	0	p	p	p	0	0

We still have no negative numbers in the last row, so we multiply each of the rows above the dotted line by $-p$ and add all of them to the last row to obtain

x_1	x_2	x_3	x_4	x_5	x_6	a	b	c	m'	
(10)	2	1	-1	0	0	1	0	0	0	50
4	3	6	0	-1	0	0	1	0	0	60
6	4	4	0	0	-1	0	0	1	0	65
80	20	15	0	0	0	0	0	0	1	0
$-20p$	$-9p$	$-11p$	p	p	p	0	0	0	0	$-175p$

We now proceed to apply the simplex method. The most negative entry in the last row is $-20p$, so the first pivot entry is 10. Completing the elimination process, we obtain

x_1	x_2	x_3	x_4	x_5	x_6	a	b	c	m'	
1	$\frac{1}{5}$	$\frac{1}{10}$	$\frac{-1}{10}$	0	0					5
0	$\frac{11}{5}$	$\frac{56}{10}$	$\frac{4}{10}$	-1	0					40
0	$\frac{14}{5}$	$\frac{34}{10}$	$\frac{6}{10}$	0	-1					35
0	4	7	8	0	0					-400
0	$-5p$	$-9p$	$-p$	p	p					$-75p$

Note that entries for columns a, b, c, and m' have not been entered or computed. This is due to the fact that the m' column remains unchanged throughout the elimination process, while the artificial variables a, b, and c will eventually disappear, so the values in their columns will not be needed.

The most negative entry in the last row is now $-9p$, so the new pivot entry is 56/10. Completing the elimination process, we obtain

x_1	x_2	x_3	x_4	x_5	x_6	a	b	c	m'	
1	$\dfrac{9}{56}$	0	$\dfrac{-3}{28}$	$\dfrac{1}{56}$	0					$\dfrac{30}{7}$
0	$\dfrac{11}{28}$	1	$\dfrac{1}{14}$	$\dfrac{-5}{28}$	0					$\dfrac{50}{7}$
0	$\left(\dfrac{41}{28}\right)$	0	$\dfrac{5}{14}$	$\dfrac{17}{28}$	-1					$\dfrac{75}{7}$
0	$\dfrac{5}{4}$	0	$\dfrac{15}{2}$	$\dfrac{5}{4}$	0					-450
0	$\dfrac{-41p}{28}$	0	$\dfrac{-5p}{14}$	$\dfrac{-17p}{28}$	p					$\dfrac{-75p}{7}$

The most negative entry in the last row is now $-41p/28$. Comparing quotients, we have

$$\frac{30}{7} \div \frac{9}{56} = \frac{80}{3} = 26\frac{2}{3}$$

$$\frac{50}{7} \div \frac{11}{28} = \frac{200}{11} = 18\frac{2}{11}$$

and

$$\frac{75}{7} \div \frac{41}{28} = \frac{300}{41} = 7\frac{13}{41}$$

so 41/28 is the new pivot entry. Completing the elimination process, we obtain

x_1	x_2	x_3	x_4	x_5	x_6	a	b	c	m'	
1	0	0	$\dfrac{-6}{41}$	$\dfrac{-12}{287}$	$\dfrac{9}{82}$			0		$\dfrac{255}{82} = x_1$
0	0	1	$\dfrac{1}{41}$	$\dfrac{98}{287}$	$\dfrac{11}{41}$			0		$\dfrac{175}{41} = x_3$
0	1	0	$\dfrac{10}{41}$	$\dfrac{17}{41}$	$\dfrac{-28}{41}$			0		$\dfrac{300}{41} = x_2$
0	0	0	$\dfrac{295}{41}$	$\dfrac{30}{41}$	$\dfrac{35}{41}$			1		$\dfrac{-18,825}{41} = m'$
0	0	0	0	0	0			0		0

There are no more negative entries in the last row, so we have arrived at an optimal solution. Since there are no longer any values containing p, the solution for m' is also the solution for m. Thus, our solution is

$$x_1 = \frac{255}{82} = 3\frac{9}{82}$$

$$x_2 = \frac{300}{41} = 7\frac{13}{41}$$

$$x_3 = \frac{175}{41} = 4\frac{11}{41}$$

and

$$m = \frac{-18,825}{41} = -459\frac{6}{41}.$$

Since $m = -C$, then

$$C = \$4.59\frac{6}{41}$$

is the minimum cost that fulfills the requirements.

Exercise Set 7.2
1. Use the simplex method to solve Problem 3 of Exercise Set 7.1.
2. Use the simplex method to solve Problem 4 of Exercise Set 7.1.
3. Use the simplex method to solve linear programs with the following constraints in addition to the nonnegative constraints on each variable.
    (a) Maximize $f = 2x_1 + x_2$ under the constraints
    $$5x_1 + 2x_2 \leq 14$$
    $$7x_1 + x_2 \leq 10.$$
    (b) Maximize $f = 5x_1 + 6x_2 + 4x_3$ subject to the constraints
    $$x_1 + x_2 + x_3 \leq 27$$
    $$2x_1 \quad\quad - x_3 \leq 0$$
    $$-x_1 + x_2 - x_3 \leq 0.$$

(c) Maximize $f = 3x_1 + 2x_2 - x_3$ subject to the constraints
$$5x_1 + 2x_3 \leqslant 19$$
$$-x_1 + x_2 + 2x_3 \leqslant 4$$
$$-8x_1 + 4x_2 + 28 \leqslant 0.$$
(d) Minimize $f = x_1 - 2x_2$ subject to the constraints
$$x_1 + \tfrac{3}{10}x_2 \leqslant 3$$
$$2x_1 + 5x_2 \leqslant 10$$
$$x_1 - x_2 \leqslant 3.$$
(e) Minimize $f = 5x_1 + 6x_2$ subject to the constraints
$$x_1 + x_2 \geqslant 3$$
$$x_1 + 2x_2 \geqslant 4.$$

4. A factory produces two types of ball-point pens, A and B, on two machines. Type A sells for a profit of 4¢ each and can be produced in 3 min on the first machine and in 2 min on the second. Type B sells for a profit of 3¢ each and can be produced in 2 min on the first machine and in 5 min on the second machine. Both machines are available for use only 60 hr per week. How many pens of each type should be produced per week to maximize profits?

5. A men's clothing factory produces pants, which bring a profit of $10.00 each; sport coats, which bring a profit of $25.00 each; and topcoats, which bring a profit of $35.00 each, and has three employees. It takes employee A 3 hr to produce a pair of pants, 5 hr to produce a sport coat, and 4 hr to produce a topcoat. It takes employee B 2 hr to produce a pair of pants, 4 hr to produce a sport coat, and 8 hr to produce a topcoat. It takes employee C 1 hr to produce a pair of pants, 3 hr to produce a sport coat, and 5 hr to produce a topcoat. Employees A and B can work 50 hr each per week, while employee C works only 20 hr per week. How many of each item should be produced per week to maximize profits?

6. A man who sharpens tools for a living finds that his business consists of sharpening lawn mowers, sharpening kitchen knives, and sharpening scissors. He finds that he averages 8 min per lawn mower for which he receives 60¢, 4 min per kitchen knife for which he receives 40¢, and 6 min per pair of scissors for which he receives 50¢. He works 7 hr per day, no more than 5 of which can be spent sharpening scissors and knives, due to his need for

variety in his work. How many articles of each type should he sharpen per day to maximize profits?

7.   In balancing a ration for his pigs, a farmer decides to try a mixture that requires at least 25 units of wheat and 24 units of rye per day. He finds, however, that he can buy neither wheat nor rye that is clean. He can buy "wheat," which is actually 4 units of wheat and 1 unit of rye for $3.00 per sack, and what is labeled "rye," which is actually 5 units of rye and 1 unit of wheat, for $1.00 per sack. He cannot buy partial sacks. How many sacks of each should he buy per day to keep his cost at a minimum?

7.4 AN INTRODUCTION TO THE THEORY OF GAMES

A firm foundation for the theory of games was first laid by John von Neumann, who proved the fundamental theory of game theory, the *minimax theorem,* in 1928. Game theory was then developed as a method for analyzing "games," which can be competition situations in warfare, economics, business, agriculture, and many other areas.

We shall consider only those games that are played by only two "players," which may be the buyer and the seller, the farmer and nature, or opposing armies. Since an exhaustive study of the game theory that has been developed is far beyond the level of this book, we include only a brief introduction to the topic here. While the games that we discuss are quite elementary, the concepts and methods can be extended to more meaningful games.

Let us consider games with two players, each of whom can make any one of several moves, and for which a winner is determined and a payoff made. Such games can be expressed and analyzed in matrix form. For example, consider a game of matching pennies where each player shows a penny; player R keeps both pennies if they both show heads or both show tails, and player C keeps both coins if one shows a head while the other shows a tail. This game can be illustrated by the table

		Player C	
		H	T
Player R	H	1	-1
	T	-1	1

in which the matrix

$$\begin{bmatrix} 1 & -1 \\ -1 & 1 \end{bmatrix}$$

is called the *payoff matrix* for player R; that is, each entry indicates the payoff to player R for the given outcome. Thus, negative entries indicate a payoff from player R to player C. That is, on each payoff the amount *won* by one player is the exact amount *lost* by the other player. Thus, such games are also called *zero-sum* games.

If the above game is played a great number of times, the long-run expectancy of R is 0, provided that neither player lets his opponent know in advance the outcome of his play, and that each will play *heads* or *tails* with equal frequency. If, however, the frequency of heads and tails (called the *strategy*) is changed, then so is the expected outcome. To determine the expected outcome in such cases, we let **A** be the payoff matrix, the row vector **P** be the strategy of R, and the column vector **Q** be the strategy of C. The expectation of R is computed as the matrix product

$$R = \mathbf{PAQ}.$$

Example

(i) Suppose that player R in the coin game adopts the strategy of playing heads $\frac{2}{3}$ of the time and tails the other $\frac{1}{3}$ of the time, while player C adopts the strategy of playing heads $\frac{3}{4}$ of the time and tails $\frac{1}{4}$ of the time; that is,

$$\mathbf{P} = [\tfrac{2}{3}, \tfrac{1}{3}] \quad \text{and} \quad \mathbf{Q} = \begin{bmatrix} \tfrac{3}{4} \\ \tfrac{1}{4} \end{bmatrix}.$$

The expectation of R is computed as

$$R = \mathbf{PAQ} = [\tfrac{2}{3}, \tfrac{1}{3}] \begin{bmatrix} 1 & -1 \\ -1 & 1 \end{bmatrix} \begin{bmatrix} \tfrac{3}{4} \\ \tfrac{1}{4} \end{bmatrix}$$

$$= [\tfrac{1}{3}, -\tfrac{1}{3}] \begin{bmatrix} \tfrac{3}{4} \\ \tfrac{1}{4} \end{bmatrix}$$

$$= \tfrac{1}{6}.$$

Thus, R can expect to win an average of $\frac{1}{6}$¢ per game over the long run.

(ii) If R adopts the strategy

$$\mathbf{P} = [\tfrac{2}{3}, \tfrac{1}{3}]$$

and C adopts the strategy $\begin{bmatrix} \frac{1}{4} \\ \frac{3}{4} \end{bmatrix}$, the expectation of R is

$$\mathbf{R} = \mathbf{PAQ} = [\tfrac{2}{3}, \tfrac{1}{3}]\begin{bmatrix} 1 & -1 \\ -1 & 1 \end{bmatrix}\begin{bmatrix} \frac{1}{4} \\ \frac{3}{4} \end{bmatrix}$$

$$= [\tfrac{1}{3}, -\tfrac{1}{3}]\begin{bmatrix} \frac{1}{4} \\ \frac{3}{4} \end{bmatrix}$$

$$= -\tfrac{1}{6}.$$

Thus, R can expect to lose an average of $\frac{1}{6}$¢ per game over the long run.

From the previous example, we see that if one player can determine the strategy of the other player, he can often adopt a strategy of his own that will cause him to obtain a greater payoff. Now let us consider the case in which the players are equally clever and both are able to determine the strategy of the other. For example, consider the game in which each player may select a card labeled 1, 2, or 3. If both cards selected are odd numbers, then player R wins the sum of the numbers in dollars. If one card is even and the other is odd, then player C wins the sum in dollars. If both cards are even, then no payoff is made. The payoff matrix is the following:

$$\begin{array}{c c} & \begin{array}{c c c} 1 & 2 & 3 \end{array} \\ \begin{array}{c} 1 \\ 2 \\ 3 \end{array} & \begin{bmatrix} 2 & -3 & 4 \\ -3 & 0 & -5 \\ 4 & -5 & 6 \end{bmatrix} \end{array}.$$

In determining a strategy, player R notices that his greatest possible payoff is $6.00 in any one game, and that the 6 is in the third row; that is, it can be obtained only by selecting card 3. At first, this seems like his best choice, but then he also notices that with this selection, he also stands to obtain his greatest possible loss, -5. Being conservative, he decides to select the choice that could bring him the least possible loss. By inspecting the payoff matrix, he observes that the greatest possible loss in the first row is -3, while the greatest possible loss in both the second and the third rows is -5. Thus, his best choice to minimize his maximum loss, thus maximize his profits, is the first row, that is, card 1. By the same reasoning, player C notices that he could lose at most \$4.00 by selecting the first column, \$0.00 by selecting the second column, and \$6.00 by selecting the third column. Thus, his best choice to maximize profits is column 2, that is, card 2. Now, if both players make their *best* move, then player R chooses card 1, and player C chooses card 2. Thus, the payoff is found in the first row and second column

$$
\begin{array}{c c c}
 & 1 & 2 & 3 \\
\begin{array}{c} 1 \\ 2 \\ 3 \end{array} &
\left[\begin{array}{c c c}
2 & \enclose{circle}{-3} & 4 \\
-3 & 0 & -5 \\
4 & -5 & 6
\end{array}\right].
\end{array}
$$

Thus, player C is paid \$3.00 by player R. Since R does not relish the idea of losing \$3.00 each play, he realizes that he should vary his play and use a strategy that includes more than the first row. Such a strategy is called a *mixed strategy* as opposed to a *pure strategy* of playing the same row or the same column on every play. Thus, a pure strategy is a *unit* row vector or a *unit* column vector; that is, one entry is 1, and all other entries are 0.

The best possible strategy for a player to make, whether pure or mixed, is called his *optimal strategy*. If both optimal strategies are also pure strategies, then the game is called *strictly determined*. Since an optimal strategy is not always immediately evident, we use another method for determining whether or not a game is strictly determined.

definition 7.1

A matrix game is *strictly determined* if and only if the matrix has an entry that is simultaneously a maximum of row minimums and a minimum of column maximums; that is, it is a minimum in its row and a maximum in its column. Such an entry is called a *saddle point*.

Example

(i)   The matrix game denoted by

$$\begin{bmatrix} -4 & 5 & -3 \\ 6 & 10 & 4 \\ -7 & -9 & 2 \end{bmatrix}$$

has 4 as a saddle point, so is strictly determined. The optimal strategies are the pure strategies

$$P = [0,1,0] \quad \text{and} \quad Q = \begin{bmatrix} 0 \\ 0 \\ 1 \end{bmatrix}.$$

(ii) The matrix game denoted by

$$\begin{bmatrix} 4 & -5 \\ -1 & 2 \end{bmatrix}$$

has -1 as its maximum of row minimums and 2 as its minimum of column maximums. Thus, it has no saddle point and is not strictly determined.

(iii) The matrix game denoted by

$$\begin{bmatrix} -5 & -2 & 4 & 3 \\ -2 & 1 & 0 & 8 \end{bmatrix}$$

has -2 as a saddle point, so is a strictly determined game. The optimal strategies are

$$P = [0,1] \quad \text{and} \quad Q = \begin{bmatrix} 1 \\ 0 \\ 0 \\ 0 \end{bmatrix}.$$

Exercise Set 7.3

1.   Given the following matrix games with **P** as the strategy for player R (represented by the rows) and **Q** as the strategy for C (represented by the columns). Find the expectancy of R.

(a) $A = \begin{bmatrix} 12 & -8 \\ -4 & 6 \end{bmatrix}$ $P = [1,0]$ $Q = \begin{bmatrix} 0 \\ 1 \end{bmatrix}$

(b) $A = \begin{bmatrix} 2 & -3 \\ -4 & 6 \end{bmatrix}$ $P = [\frac{1}{4},\frac{3}{4}]$ $Q = \begin{bmatrix} \frac{1}{2} \\ \frac{1}{2} \end{bmatrix}$

(c) $A = \begin{bmatrix} 2 & -3 \\ -4 & 6 \end{bmatrix}$ $P = [\frac{2}{3},\frac{1}{3}]$ $Q = \begin{bmatrix} \frac{3}{5} \\ \frac{2}{5} \end{bmatrix}$

(d) $A = \begin{bmatrix} -2 & 0 & 3 \\ 5 & 2 & -4 \end{bmatrix}$ $P = [\frac{2}{3},\frac{1}{3}]$ $Q = \begin{bmatrix} \frac{1}{3} \\ 0 \\ \frac{2}{3} \end{bmatrix}$

(e) $A = \begin{bmatrix} 0 & 5 & 6 \\ -5 & 0 & 7 \\ -6 & -7 & 0 \end{bmatrix}$ $P = [\frac{2}{3},\frac{1}{3},0]$ $Q = \begin{bmatrix} 1 \\ 0 \\ 0 \end{bmatrix}$

2. Identify each of the following matrix games as strictly determined or not strictly determined. For those that are strictly determined, identify the saddle point and the optimal strategy for both rows and columns.

(a) $\begin{bmatrix} -5 & 2 \\ -4 & 3 \end{bmatrix}$

(d) $\begin{bmatrix} 0 & -5 & 2 \\ 4 & 7 & 3 \\ -5 & -4 & -3 \end{bmatrix}$

(b) $\begin{bmatrix} 2 & -1 & 3 \\ -1 & 2 & 1 \end{bmatrix}$

(e) $\begin{bmatrix} 5 & -15 & 2 \\ -1 & 3 & -2 \end{bmatrix}$

(c) $\begin{bmatrix} -4 & 0 & 5 \\ 2 & -5 & 7 \end{bmatrix}$

(f) $\begin{bmatrix} 2 & -4 & 6 \\ 6 & -2 & 4 \end{bmatrix}$

3. Construct a payoff matrix to describe each of the following matrix games. Also, propose three distinct strategies for each (for both playing) and determine the expectancies of R for each strategy.

(a) The coin-matching game in which player R receives a payoff of $1 if the coins show the same, and player C receives a payoff of $1 if they do not show the same.

(b) A game in which each player shows one side of a three-sided dice. If the sum of the dice is even, then player R receives a payoff of their sum in dollars. If their sum is odd, then player C receives their product in dollars.

7.5 2 × 2 MATRIX GAMES

We define the value of a matrix game **A** with optimal strategies **P′** and **Q′** to be the matrix product

$$v = \mathbf{P'AQ'}.$$

We say that a game is *fair* if its value is 0. Note that the value of a game is dependent upon its optimal strategies. Since optimal strategies are not always obvious, we limit our discussion in this section to the solving of 2 × 2 matrix games.

First, consider the case of a 2 × 2 matrix game that is strictly determined. In such cases, the optimal strategies must be one of the following:

$$[1,0] \quad \text{and} \quad \begin{bmatrix} 1 \\ 0 \end{bmatrix}$$

$$[1,0] \quad \text{and} \quad \begin{bmatrix} 0 \\ 1 \end{bmatrix}$$

$$[0,1] \quad \text{and} \quad \begin{bmatrix} 1 \\ 0 \end{bmatrix}$$

or

$$[0,1] \quad \text{and} \quad \begin{bmatrix} 0 \\ 1 \end{bmatrix}.$$

In each case, the value v of the game is the saddle point itself.

Now we consider the case in which the matrix game is nonstrictly determined. We can use the methods of the previous section to identify a matrix game as strictly or nonstrictly determined, or we can use the following theorem.

theorem 7.1

The matrix game

$$\mathbf{A} = \begin{bmatrix} a & b \\ c & d \end{bmatrix}$$

is nonstrictly determined if and only if both entries on one of the diagonals are greater than each entry on the other diagonal; that is,

$$a, d > b \quad \text{and} \quad a, d > c$$

or

$$b, c > a \quad \text{and} \quad b, c > d.$$

Example

(i) The matrix game

$$\begin{bmatrix} 4 & -5 \\ -1 & 2 \end{bmatrix}$$

is nonstrictly determined, because both 4 and 2 are greater than -1 and greater than -5.

(ii) The matrix game

$$\begin{bmatrix} 3 & -2 \\ 4 & 2 \end{bmatrix}$$

is strictly determined, because $4 > 3$ and $4 > 2$, but $-2 < 3$ and $-2 < 2$. Also, the matrix has 2 as a saddle point.

In the case of a nonstrictly determined game, mixed strategies appear as optimal strategies, but these cannot always be found by inspection. Thus, we use the following theorem.

theorem 7.2

The strategies

$$\mathbf{P} = [p_1, p_2] \quad \text{and} \quad \mathbf{Q} = \begin{bmatrix} q_1 \\ q_2 \end{bmatrix}$$

of the nonstrictly determined game

$$\mathbf{A} = \begin{bmatrix} a & b \\ c & d \end{bmatrix}$$

are optimal strategies of R and C, respectively, where

$$p_1 = \frac{d - c}{(a + d) - (b + c)} \qquad p_2 = \frac{a - b}{(a + d) - (b + c)}$$

and

$$q_1 = \frac{d-b}{(a+d)-(b+c)} \qquad q_2 = \frac{a-c}{(a+d)-(b+c)}.$$

A direct result of this theorem and the equation

$$v = \mathbf{P'AQ'}$$

is the equation

$$v = \frac{ad-bc}{(a+d)-(b+c)}$$

for the value of the nonstrictly determined matrix game

$$\begin{bmatrix} a & b \\ c & d \end{bmatrix}.$$

Example

(i)   Given the nonstrictly determined matrix game

$$\begin{bmatrix} 1 & -1 \\ -1 & 1 \end{bmatrix}$$

we obtain the optimal strategies

$$\mathbf{P'} = \left[\frac{d-c}{(a+d)-(b+c)} \;,\; \frac{a-b}{(a+d)-(b+c)} \right]$$

$$= \left[\frac{1-(-1)}{2-(-2)} \;,\; \frac{1-(-1)}{2-(-2)} \right]$$

$$= \left[\frac{2}{4}, \frac{2}{4} \right] = \left[\frac{1}{2}, \frac{1}{2} \right]$$

$$\mathbf{Q'} = \begin{bmatrix} \dfrac{d-b}{(a+d)-(b+c)} \\[2ex] \dfrac{a-c}{(a+d)-(b+c)} \end{bmatrix} = \begin{bmatrix} \dfrac{1-(-1)}{2-(-2)} \\[2ex] \dfrac{1-(-1)}{2-(-2)} \end{bmatrix} = \begin{bmatrix} \dfrac{2}{4} \\[1ex] \dfrac{2}{4} \end{bmatrix} = \begin{bmatrix} \dfrac{1}{2} \\[1ex] \dfrac{1}{2} \end{bmatrix}$$

and the value

$$v = \frac{ad - bc}{(a + d) - (b + c)} = \frac{1 - 1}{4} = \frac{0}{4} = 0.$$

Thus, the game is fair.

(ii) The nonstrictly determined matrix game

$$\begin{bmatrix} 4 & -3 \\ -3 & 2 \end{bmatrix}$$

has optimal strategies

$$\mathbf{P'} = \begin{bmatrix} \dfrac{2 - (-3)}{6 - (-6)}, \dfrac{4 - (-3)}{6 - (-6)} \end{bmatrix} = \begin{bmatrix} \dfrac{5}{12} & , & \dfrac{7}{12} \end{bmatrix}$$

$$\mathbf{Q'} = \begin{bmatrix} \dfrac{2 - (-3)}{6 - (-6)} \\ \dfrac{4 - (-3)}{6 - (-6)} \end{bmatrix} = \begin{bmatrix} \dfrac{5}{12} \\ \dfrac{7}{12} \end{bmatrix}$$

and value

$$v = \frac{8 - 9}{6 - (-6)} = \frac{-1}{12}.$$

Thus, the game is not fair; it favors player C, who plays the columns.

(iii) The nonstrictly determined matrix game

$$\begin{bmatrix} 2 & -3 \\ -4 & 6 \end{bmatrix}$$

has optimal strategies

$$\mathbf{P'} = \begin{bmatrix} \dfrac{6 - (-4)}{8 - (-7)}, \dfrac{2 - (-3)}{8 - (-7)} \end{bmatrix} = \begin{bmatrix} \dfrac{10}{15} & , & \dfrac{5}{15} \end{bmatrix} = \begin{bmatrix} \dfrac{2}{3} & , & \dfrac{1}{3} \end{bmatrix}$$

$$\mathbf{Q'} = \begin{bmatrix} \dfrac{6 - (-3)}{8 - (-7)} \\ \dfrac{2 - (-4)}{8 - (-7)} \end{bmatrix} = \begin{bmatrix} \dfrac{9}{15} \\ \dfrac{6}{15} \end{bmatrix} = \begin{bmatrix} \dfrac{3}{5} \\ \dfrac{2}{5} \end{bmatrix}$$

and value

$$v = \frac{12 - 12}{15} = \frac{0}{15} = 0.$$

Thus, the game is fair.

Note that the only way the value of a nonstrictly determined game

$$\begin{bmatrix} a & b \\ c & d \end{bmatrix}$$

can be 0 is for the numerator of the fraction

$$\frac{ad - bc}{(a + d) - (b + c)}$$

to be 0. Thus, the game is fair only if $ad - bc = 0$; that is, $ad = bc$.

Exercise Set 7.4

1. Identify each of the following matrix games as strictly or non-strictly determined and find the optimal strategies of each.

(a) $\begin{bmatrix} -1 & 2 \\ 4 & 3 \end{bmatrix}$

(g) $\begin{bmatrix} -2 & 3 \\ 4 & -6 \end{bmatrix}$

(b) $\begin{bmatrix} 2 & 0 \\ -1 & 3 \end{bmatrix}$

(h) $\begin{bmatrix} 0 & 2 \\ -3 & 4 \end{bmatrix}$

(c) $\begin{bmatrix} 0 & 3 \\ -4 & 0 \end{bmatrix}$

(i) $\begin{bmatrix} -2 & 5 \\ 4 & 3 \end{bmatrix}$

(d) $\begin{bmatrix} 2 & -3 \\ -3 & 4 \end{bmatrix}$

(j) $\begin{bmatrix} 0 & -7 \\ 0 & 7 \end{bmatrix}$

(e) $\begin{bmatrix} -5 & 10 \\ 2 & -4 \end{bmatrix}$

(k) $\begin{bmatrix} 4 & -10 \\ -2 & 5 \end{bmatrix}$

(f) $\begin{bmatrix} 3 & -1 \\ -4 & 2 \end{bmatrix}$

(l) $\begin{bmatrix} -4 & 10 \\ -2 & 5 \end{bmatrix}$

2. Find the value of each matrix game in Problem 1.
3. Label each game in Problem 1 as favoring player R, favoring player C, or fair.
4. Set up each of the following games as a matrix game and analyze it as strictly determined or nonstrictly determined and find the opti-

mal strategies for both players; find the value of the game; and identify it as fair or favoring a particular player.

(a)   The coin-matching game.

(b)   A game in which each player may select either a 1 or a 2. If the same number is selected by both players, then player R receives a payoff of their product in dollars. If different numbers are selected, then player C receives a payoff of their sum in dollars.

(c)   A football game in which the offensive team has only two offenses, run or pass. The defensive team knows that the offensive has only two plays, so only two defenses are incorporated – a running defense and a passing defense. If the offensive passes against the running defense, the average gain will be 8 yd; if a running play is called against the pass defense, the average gain will be 5 yd; if the defense matches the play, then there is no gain or loss.

(d)   Miss Jones, a student at Podunk College, must take a particular math class, which is taught twice per day by each of two professors – Dr. Morgan and Dr. Aften. She knows that Dr. Morgan is much the better teacher in the morning and that Dr. Aften is much the better teacher in the afternoon. She also knows that the registration system is such that she can select the hour in which she takes the class, but the math secretary, whose boyfriend Miss Jones just stole, assigns each student to a particular professor. Each must make her decision without knowing the decision made by the other. Consider the payoff for Miss Jones to be 1 if she gets the more desirable instructor at either time, and -1 if not.

7.6 $m \times n$ MATRIX GAMES AND LINEAR PROGRAMMING

The techniques used for solving 2×2 matrix games can often (but not usually) be used to solve $m \times n$ matrix games. For example, consider the matrix game

$$\mathbf{A} = \begin{bmatrix} 5 & -2 & 2 \\ -4 & 3 & -5 \\ 3 & -3 & 1 \end{bmatrix}.$$

Note in this matrix that every entry of the first row is *greater* than the corresponding entry of the third row. We say that the first row *dominates* the third row; that is, in searching for an optimal strategy, the row player would *never* choose row 3. Thus, in trying to obtain optimal solutions, the problem is reduced to the matrix game

$$\begin{bmatrix} 5 & -2 & 2 \\ -4 & 3 & -5 \end{bmatrix}.$$

Now consider the game from the column player's point of view. Every entry in column 3 is *less* than (greater than for player C) the corresponding entry in column 1. Thus, column 3 dominates column 1, and column 1 can be eliminated from consideration in the optimal strategy. We, therefore, have reduced the problem to the matrix game

$$\begin{bmatrix} -2 & 2 \\ 3 & -5 \end{bmatrix}$$

whose optimal strategies $\mathbf{P}'$ and $\mathbf{Q}'$ will also be the optimal strategies of the original game $\mathbf{A}$. The value of $\mathbf{A}$ can be computed by the equation

$$v = \mathbf{P}'\mathbf{A}\mathbf{Q}'$$

which we have already been using for computing value.

Now let us consider an $m \times n$ matrix game that cannot be reduced to an equivalent 2×2 matrix game by the method just discussed. We need to employ some other procedure for obtaining optimal strategies. The first step of the procedure that we shall use is to observe whether or not the matrix contains any negative entries, and if so, to add a constant term K to every entry, such that all entries will be positive. This procedure has no effect on the strategies being employed, since it merely adds $K > 0$ to each expectancy, including v, regardless of the strategy being used. Now let us consider any strategies

$$\mathbf{P} = [p_1, p_2, \ldots, p_m] \quad \text{and} \quad \mathbf{Q} = \begin{bmatrix} q_1 \\ q_2 \\ \cdot \\ \cdot \\ \cdot \\ q_n \end{bmatrix}$$

where the value of the matrix game $\mathbf{A}$ is v. Since v is the value of $\mathbf{A}$, the product of $\mathbf{P}$ with any column of $\mathbf{A}$ must be greater than or equal to v, and the product of any row of $\mathbf{A}$ with $\mathbf{Q}$ must be less than or equal to v; that is, if

$$\mathbf{A} = \begin{bmatrix} a_{11} & a_{12} & \cdots & a_{1n} \\ a_{21} & a_{22} & \cdots & a_{2n} \\ \cdot & \cdot & \cdot & \cdot \\ \cdot & \cdot & \cdot & \cdot \\ \cdot & \cdot & \cdot & \cdot \\ a_{m1} & a_{m2} & \cdots & a_{mn} \end{bmatrix}$$

then

$$p_1 a_{11} + p_2 a_{21} + p_3 a_{31} + \cdots + p_m a_{m1} \geq v$$
$$p_1 a_{12} + p_2 a_{22} + p_3 a_{32} + \cdots + p_m a_{m2} \geq v$$
$$\cdot \qquad \cdot \qquad \cdot \qquad \qquad \cdot$$
$$\cdot \qquad \cdot \qquad \cdot \qquad \qquad \cdot$$
$$\cdot \qquad \cdot \qquad \cdot \qquad \qquad \cdot$$
$$p_m a_{1n} + p_2 a_{2n} + p_3 a_{3n} + \cdots + p_m a_{mn} \geq v$$

and

$$a_{11} q_1 + a_{12} q_2 + a_{13} q_3 + \cdots + a_{1n} q_n \leq v$$
$$a_{21} q_1 + a_{22} q_2 + a_{23} q_3 + \cdots + a_{2n} q_n \leq v$$
$$\cdot \qquad \cdot \qquad \cdot \qquad \qquad \cdot$$
$$\cdot \qquad \cdot \qquad \cdot \qquad \qquad \cdot$$
$$\cdot \qquad \cdot \qquad \cdot \qquad \qquad \cdot$$
$$a_{m1} q_1 + a_{m2} q_2 + a_{m3} q_3 + \cdots + a_{mn} q_n \leq v.$$

Remember that the optimal strategy $\mathbf{P}'$ is a minimum such that $\mathbf{P}'\mathbf{A} \geq v$, and the optimal strategy $\mathbf{Q}'$ is a maximum such that $\mathbf{A}\mathbf{Q}' \leq v$. Thus, we can obtain such optimal strategies through linear programming with the previous sets of inequalities as constraints. To do so, we let $y_j = p_j/v$ and $x_i = q_i/v$ and proceed to maximize $f = x_1 + x_2 + \cdots + x_n$, subject to the constraints

$$a_{11} x_1 + a_{12} x_2 + a_{13} x_3 + \cdots + a_{1n} x_n \leq 1$$
$$a_{21} x_1 + a_{22} x_2 + a_{23} x_3 + \cdots + a_{2n} x_n \leq 1$$
$$\cdot \qquad \cdot \qquad \cdot \qquad \qquad \cdot \qquad \cdot$$
$$\cdot \qquad \cdot \qquad \cdot \qquad \qquad \cdot \qquad \cdot$$
$$\cdot \qquad \cdot \qquad \cdot \qquad \qquad \cdot \qquad \cdot$$
$$a_{m1} x_1 + a_{m2} x_2 + a_{m3} x_3 + \cdots + a_{mn} x_n \leq 1$$

and to minimize

$$g = y_1 + y_2 + \cdots + y_m$$

subject to the constraints

$$a_{11}y_1 + a_{21}y_2 + a_{31}y_3 + \cdots + a_{m1}y_m \geq 1$$
$$a_{12}y_1 + a_{22}y_2 + a_{32}y_3 + \cdots + a_{m2}y_m \geq 1$$
$$\vdots \qquad \vdots \qquad \vdots \qquad \qquad \vdots$$
$$a_{1n}y_1 + a_{2n}y_2 + a_{3n}y_3 + \cdots + a_{mn}y_m \geq 1.$$

We have $v = 1/M$, where M is the maximum value of f. Thus, we are able to solve for x_i and y_j for $i = 1,2,3, \ldots ,m$ and $j = 1,2,3, \ldots ,n$. Thus, we can obtain the optimal strategies through linear programming.

Example

To solve the matrix game

$$\begin{bmatrix} 0 & 2 \\ 1 & -1 \\ -2 & 3 \end{bmatrix}$$

we first add 3 to each entry to obtain

$$\begin{bmatrix} 3 & 5 \\ 4 & 2 \\ 1 & 6 \end{bmatrix}.$$

To solve for the optimal strategy

$$\mathbf{Q} = \begin{bmatrix} q_1 \\ q_2 \end{bmatrix}$$

we need to maximize the function $f = x_1 + x_2$ subject to the constraints

$$x_1 \geqslant 0$$
$$x_2 \geqslant 0$$
$$3x_1 + 5x_2 \leqslant 1$$
$$4x_1 + 2x_2 \leqslant 1$$
$$x_1 + 6x_2 \leqslant 1$$

and to minimize the function $g = y_1 + y_2 + y_3$ subject to the constraints

$$y_1 \geqslant 0$$
$$y_2 \geqslant 0$$
$$y_3 \geqslant 0$$
$$3y_1 + 4y_2 + y_3 \geqslant 1$$
$$5y_1 + 2y_2 + 6y_3 \geqslant 1.$$

To maximize the first program, we can use the simplex method or solve geometrically, since the graph is in two dimensions only. We obtain a solution $x_1 = \frac{3}{14}$ and $x_2 = \frac{1}{14}$ with maximum value $M = f = x_1 + x_2 = \frac{3}{14} + \frac{1}{14} = \frac{4}{14}$. Thus, $v = 1/M = \frac{14}{4}$, and the entries q_1 and q_2 can be obtained by

$$q_1 = x_1 v = \left(\frac{3}{14}\right)\left(\frac{14}{4}\right) = \frac{3}{4}$$

$$q_2 = x_2 v = \left(\frac{1}{14}\right)\left(\frac{14}{4}\right) = \frac{1}{4}.$$

Therefore, the optimal strategy for playing the columns is

$$\mathbf{Q}' = \begin{bmatrix} \frac{3}{4} \\ \frac{1}{4} \end{bmatrix}.$$

$\mathbf{P}'$ can now be found by the simplex method, in which g is minimized by maximizing $-g$. This is left as an exercise.

Note that the value v, which is used in computing $\mathbf{P}'$ and $\mathbf{Q}'$, is not necessarily the value of the original game. In this case, the value of the original game is $v - 3$, since 3 was added to each entry. In each case where K is added to each entry, the value of the original game is $v - K$, where $v = 1/M$.

Exercise Set 7.5

1. Reduce the following matrix games as far as possible by eliminating all dominated strategies (both rows and columns).

    (a) $\begin{bmatrix} 1 & 0 & 2 \\ 0 & 3 & 1 \end{bmatrix}$

    (b) $\begin{bmatrix} 0 & -2 & 1 \\ 7 & 6 & -3 \\ -1 & -3 & -7 \end{bmatrix}$

    (c) $\begin{bmatrix} -5 & -3 & 4 \\ 5 & 3 & 6 \\ 2 & -5 & -3 \end{bmatrix}$

    (d) $\begin{bmatrix} 4 & 1 \\ 1 & 7 \\ 12 & 3 \end{bmatrix}$

    (e) $\begin{bmatrix} 2 & -3 \\ -5 & 4 \\ 4 & 1 \end{bmatrix}$

    (f) $\begin{bmatrix} 4 & -3 & -7 \\ -2 & 1 & 0 \\ 5 & 0 & -5 \end{bmatrix}$

    (g) $\begin{bmatrix} 3 & -2 \\ -6 & 4 \\ 5 & 3 \end{bmatrix}$

    (h) $\begin{bmatrix} 3 & 2 & 0 \\ 6 & 4 & 3 \\ 0 & 2 & 3 \end{bmatrix}$

    (i) $\begin{bmatrix} -3 & 2 & 5 \\ -4 & -1 & -1 \\ -10 & 1 & 3 \end{bmatrix}$

    (j) $\begin{bmatrix} -5 & 4 & 5 \\ 4 & 1 & 3 \\ 6 & -4 & -3 \end{bmatrix}$

2. Find the optimal strategy for each player and the value of the game for each exercise in Problem 1.

3. Identify each of the games in Problem 1 as fair or unfair.

4. Find all saddle points of games in Problem 1.

5. Find P' in the example preceding this exercise set.

6. Use whatever methods seem most appropriate to find the optimal strategies and value of each of the following matrix games.

    (a) $\begin{bmatrix} 5 & 4 \\ -1 & 6 \\ -2 & 5 \end{bmatrix}$

    (b) $\begin{bmatrix} 2 & 0 \\ -1 & 10 \\ 1 & 2 \end{bmatrix}$

    (c) $\begin{bmatrix} -1 & 0 \\ 0 & -2 \\ 1 & -3 \end{bmatrix}$

    (d) $\begin{bmatrix} -1 & 1 & 3 \\ 0 & -1 & -3 \end{bmatrix}$

    (e) $\begin{bmatrix} 1 & 5 \\ 3 & 4 \\ 2 & 1 \end{bmatrix}$

    (f) $\begin{bmatrix} -1 & 4 & 1 \\ 3 & 3 & 0 \\ 0 & -1 & 3 \end{bmatrix}$

    (g) $\begin{bmatrix} 5 & -2 & -1 \\ 3 & 1 & 0 \\ -4 & 5 & -2 \end{bmatrix}$

    (h) $\begin{bmatrix} 0 & -1 \\ 3 & 1 \\ 1 & 2 \end{bmatrix}$

(i) $\begin{bmatrix} -7 & 2 \\ 2 & -1 \\ 4 & -6 \end{bmatrix}$ (j) $\begin{bmatrix} 5 & -1 & -2 & 0 \\ -1 & 0 & 0 & 0 \\ 2 & 1 & 2 & 1 \\ 5 & 1 & 3 & 1 \end{bmatrix}$

7. Illustrate each of the following as a matrix game and solve for its value and its optimal strategies.

(a) A dice game in which player R rolls one dice and receives a payoff of $1 if the number of dots is even and pays player C $1 if the number of dots is odd.

(b) An incumbent president running for reelection is considering three strategies for campaigning—ignoring his opponent by campaigning as a statesman, mildly campaigning by speaking to the issues only, and attacking vigorously through character assassination. His opponent has the same three strategies available. The president knows that he will win the election provided that he and his opponent campaign at the same level or only one degree, or strategy, apart (score 1). He also knows that he will lose the election (score −3) provided that he is either two strategies milder or two strategies more active than his opponent. This same information is available to both candidates, and neither can change the level of the campaign once it has begun.

(c) Mr. Jackson, a farmer in Aridville, must plant his crops in the fall of the year, and he can plant wheat, barley, or alfalfa on all or any portion of his farm. His dilemma stems from the fact that he must have an abundant snowfall for the alfalfa to make a crop, and only a light snowfall or better for the wheat to make a crop. The following table illustrates the dollars per acre profit to be expected under all conditions.

Crop	Snowfall		
	Heavy	Average	Light
Alfalfa	100	0	−30
Barley	80	20	−10
Wheat	50	40	30

(d) The game of "scissors, rock, and paper" in which "scissors cut paper, paper covers rock, and rock breaks scissors." Thus, each player may select scissors, rock, or paper. If they choose the same article, there is no payoff, but if one selects scissors and the other selects paper, then scissors is the winner. Likewise, if one selects scissors and the other selects rock, then rock is the winner, and if one selects rock and the other selects paper, then paper is the winner. Each winner receives a payoff of $1.

(e) Two co-eds, Sue and Nancy, have been periodically dating both Terry Galahad and Ron Ronchie. Sue takes chemistry with both men, and Nancy takes psychology with both. Each co-ed would like to have Terry drop in at her apartment tonight and knows that her behavior in class will determine whether or not he does. The problem is that both girls want Terry and neither wants Ron, yet they must treat them the same since they sit together. Each girl can employ any one of three strategies — cool, cordial, or charming. If either girl is just one strategy friendlier than the other, she knows that Terry will show up and Ron will stay home (Terry is somewhat more perceptive and needs less encouragement), which is considered a payoff of 1. If both co-eds employ the same strategy, both men stay home unable to decide which to visit, which is a score of 0. If, however, one co-ed is two strategies friendlier than the other, it is noted by both men and both show up at the apartment, which is disastrous and scored as −2. Since Ron is not too perceptive to encouragement, there is no chance of him showing up at either apartment alone.

Assume Sue and Nancy to be active opponents, each of whom feels gain according to the other's misery. Also assume that neither knows of the strategy being employed by the other.

A
GRAPHS
OF INEQUALITIES

Geometric methods of solving linear programs are somewhat dependent upon one's ability to graph linear equations and inequalities. Thus, we review such methods here.

First, consider the equation of a line in the form

$$y = mx + b$$

in which m represents the slope of the line and b represents its y intercept. The point $(0,b)$ is the point at which the line crosses the vertical axis, while the slope m is some rational number h/k, which indicates that the line rises h units for every k units of run (to the right) from any point on the line. Thus, we can obtain two points $(0,b)$ and $(k, b + h)$, which determine the graph of the line.

Example

(i)   The graph of the linear equation $y = 2x + 3$ can be determined by the point $(0,3)$ and the point 2 units above and 1 unit to the right of $(0,3)$; that is, $(0 + 1,\ 3 + 2) = (1,5)$.

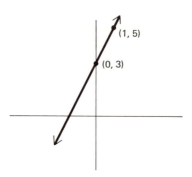

(ii) The graph of the linear equation $2x + 3y = 6$ is the graph of the equivalent equation $y = -\frac{2}{3}x + 2$. The y intercept is 2, and the slope is $-\frac{2}{3}$; so the graph can be determined by the point $(0,2)$ and the point 2 units below and 3 units to the right of $(0,2)$; that is, $(0 + 3,\ 2 - 2) = (3,0)$.

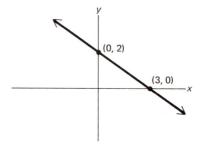

If the equation of a line cannot be expressed in the form $y = mx + b$, then the line is of the form $x = C$, which is a vertical line parallel to and C units to the right of the y axis. A horizontal line has a slope of 0, so its equation is of the form $y = b$.

Just as a linear equation can be expressed in the form $y = mx + b$, a linear inequality can be expressed in one of the forms

$$y < mx + b \qquad\quad y \leqslant mx + b$$
$$y > mx + b \quad \text{or} \quad y \geqslant mx + b.$$

The first step in the graphing of any of these inequalities is to graph the line $y = mx + b$. Now we graph $y \leqslant mx + b$ by shading the area *below* the line, and we graph $y \geqslant mx + b$ by shading the area *above* the line. We leave the line drawn solid to indicate that it is a part of the graph. To graph $y < mx + b$ and $y > mx + b$, we follow exactly the same procedure, respectively, as with $y \leqslant mx + b$ and $y \geqslant mx + b$, except that the line $y = mx + b$ is drawn as a dotted line to indicate its exclusion.

Example

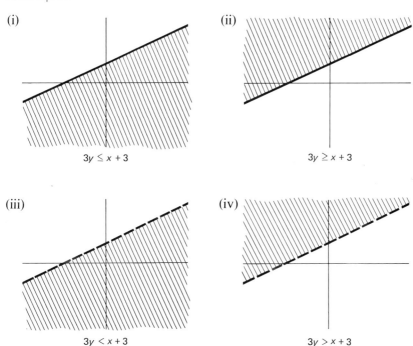

(i)

$3y \leq x + 3$

(ii)

$3y \geq x + 3$

(iii)

$3y < x + 3$

(iv)

$3y > x + 3$

When we graph the conjunction of two or more inequalities, we merely shade the area that is common to all of their graphs.

Example

To graph $x + y \leqslant 8$, $y < 4$, and $x + y \geqslant 2$, we first graph the lines $x + y = 8$, $y = 4$, and $x + y = 2$, and then shade in the area that is below $x + y = 8$, below $y = 4$, and above $x + y = 2$.

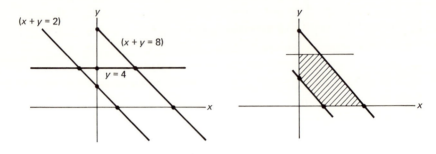

Exercise Set A.1

1. Graph the following equations and inequalities.

(a) $2x + y = 7$ (f) $2x + 3y < 0$
(b) $x - y = 3$ (g) $3x - y \geqslant -3$
(c) $3x - 2y = 9$ (h) $y \leqslant x + 3$
(d) $x + y < 5$ (i) $2x > y - 2$
(e) $x - y > 2$ (j) $5x - 2y \leqslant 7$

2. Graph the following conjunctions of inequalities.

(a) $x + y \geqslant 4$, $x \leqslant 7$, and $y \leqslant 3$
(b) $x \geqslant y$, $x \geqslant -1$, and $y \leqslant 4$
(c) $x < y + 1$, $2x + y \leqslant 10$, and $x \geqslant 0$
(d) $x < 3$, $y > 2$, and $x + y \leqslant 10$
(e) $x + y < 3$, $x - y > 3$, and $x + y > -10$

appendix
B
MATHEMATICAL
INDUCTION

The principle of *finite induction* or *mathematical induction* is a principle that allows us to prove theorems easily that would otherwise be difficult, if not impossible, to prove at this level. Intuitively, the principle of finite induction states that by proving 1) that an open sentence is true for the first element of a set, and 2) that for any element for which the sentence is true, it is also true for the next element; we, therefore, prove that the sentence is true for every element of the set. If these two conditions hold, then the sentence is true for the first element of the set; so, by the second condition, it must be true for the second element of the set. Since it is true for the second element of the set, then it must be true for the third element, and so on, to all of the elements of the set. This principle, however, is dependent upon the set *having* a first element, a second element, and a *next* element to any given element. We call this property the "well-ordered" property. Not all number systems have this property. For example, what is the *next* rational number to 1? Is it 3/2, 4/3, or 8/7, or possibly 3215/3214? Since the positive integers are well ordered, we define the principle of mathematical induction on the positive integers.

definition b.1 (mathematical induction)

If $P(n)$ is an open sentence about a positive integer n such that
1) $P(1)$ is true, and
2) $P(k + 1)$ is true whenever $P(k)$ is true for any positive integer k,

then $P(n)$ is true for every positive integer n.

Example

Let $P(n)$ be the open sentence

$$1 + 2 + 3 + \cdots + n = \frac{n(n + 1)}{2}.$$

(i) $P(1)$ is the statement

$$1 = \frac{1(1 + 1)}{2} = \frac{1(2)}{2} = 1$$

which is true.

(ii) Assume $P(k)$; that is,

$$1 + 2 + 3 + \cdots + k = \frac{k(k + 1)}{2}.$$

Adding $k + 1$ to both members, we have

$$1 + 2 + 3 + \cdots + k + (k + 1) = \frac{k(k + 1)}{2} + (k + 1).$$

Expressing the right member over a common denominator

we have

$$1 + 2 + 3 + \cdots + k + (k + 1) = \frac{k(k + 1) + 2(k + 1)}{2}.$$

Factoring $k + 1$ out to the left, we have

$$1 + 2 + 3 + \cdots + k + (k + 1) = \frac{(k + 1)(k + 2)}{2}$$

which is $P(k + 1)$. Therefore, $P(n)$ is true for all positive integers n; that is,

$$1 + 2 = \frac{2(2 + 1)}{2}$$

$$1 + 2 + 3 = \frac{3(3 + 1)}{2}$$

$$1 + 2 + 3 + 4 = \frac{4(4 + 1)}{2}$$

$$\cdot$$
$$\cdot$$
$$\cdot$$

$$1 + 2 + 3 + \cdots + n = \frac{n(n + 1)}{2}.$$

Example

Let $P(n)$ be the open sentence

$$1^2 + 2^2 + 3^2 + \cdots + n^2 = \frac{n(n + 1)(2n + 1)}{6}.$$

(i) $P(1)$ is the statement

$$1^2 = \frac{1(1 + 1)(2 + 1)}{6} = \frac{1(2)(3)}{6} = 1$$

which is true.

(ii) $P(k)$ is the statement

$$1^2 + 2^2 + 3^2 + \cdots + k^2 = \frac{k(k+1)(2k+1)}{6}.$$

Adding $(k+1)^2$ to both members, we obtain

$$1^2 + 2^2 + 3^2 + \cdots + k^2 + (k+1)^2$$
$$= \frac{k(k+1)(2k+1)}{6} + (k+1)^2.$$

Finally, by expressing the right member over a common denominator and factoring the numerator, we obtain

$$1^2 + 2^2 + 3^2 + \cdots + k^2 + (k+1)^2 = \frac{(k+1)(k+2)(2k+3)}{6}$$

which is $P(k+1)$. Therefore, $P(n)$ is true for all positive integers n.

Example (The Binomial Theorem)

For any positive integer n and real numbers a and b,

$$(a+b)^n = a^n + \binom{n}{1}(a^{n-1}b) + \binom{n}{2}(a^{n-2}b^2)$$

$$+ \cdots + \binom{n}{n-1}(ab^{n-1}) + b^n.$$

proof

A reasonable method of proving this theorem is by induction. The proof for $n = 1$ is trivial, and we have already shown that case in Chapter 3 as well as the cases for $n = 2$, $n = 3$, $n = 4$, and $n = 5$. Thus, all that remains to be proved is that if the case for $n = k$ is true, then the case for $n = k + 1$ is also true.

Assume the theorem to be true for k; that is,

$$(a+b)^k = a^k + \binom{k}{1}(a^{k-1}b) + \binom{k}{2}(a^{k-2}b^2)$$

$$+ \cdots + \binom{k}{k-1}(ab^{k-1}) + b^k.$$

Now, by multiplying each member by $(a + b)$, we obtain

$$(a + b)^{k+1} = (a^{k+1} + a^k b) + \binom{k}{1} (a^k b + a^{k-1} b^2)$$

$$+ \binom{k}{2} (a^{k-1} b^2 + a^{k-2} b^3)$$

$$+ \cdots + \binom{k}{k-1} (a^2 b^{k-1} + ab^k) + ab^k + b^{k+1}$$

Regrouping, we obtain

$$(a + b)^{k+1} = a^{k+1} + \left(1 + \binom{k}{1}\right) a^k b + \left(\binom{k}{1} + \binom{k}{2}\right) a^{k-1} b^2$$

$$+ \left(\binom{k}{2} + \binom{k}{3}\right) a^{k-2} b^3$$

$$+ \cdots + \left(\binom{k}{k-1} + 1\right) ab^k + b^{k+1}$$

in which the coefficient of $a^{(k+1)-r} b^r$ is

$$\binom{k}{r} + \binom{k}{r-1} = \frac{k!}{r!(k-r)} + \frac{k!}{(r-1)!(k-r+1)!}$$

$$= \frac{k!(k-r+1) + k!r}{r!(k-r+1)!}$$

$$= \frac{k!(k+1)}{r!(k+1-r)}$$

$$= \frac{(k+1)!}{r!(k+1-r)}$$

$$= \binom{k+1}{r}.$$

Therefore, the theorem is proved for all positive integers n.

Exercise Set B.1

Prove the following for all positive integers using mathematical induction.

1. $2 + 4 + 6 + \cdots + 2n = n(n + 1)$
2. $1 + 3 + 5 + \cdots + (2n - 1) = n^2$
3. $4 + 8 + 12 + \cdots + 4n = 2n(n + 1)$
4. $1 + 5 + 9 + \cdots + (4n - 3) = n(2n - 1)$
5. $1^3 + 2^3 + 3^3 + \cdots + n^3 = \dfrac{n^2(n + 1)^2}{4}$
6. $1^4 + 2^4 + 3^4 + \cdots + n^4 = \dfrac{n(n + 1)(2n + 1)(3n^2 + 3n - 1)}{30}$
7. $3 + 6 + 9 + \cdots + 3n = \dfrac{3n(n + 1)}{2}$
8. $3 \cdot 6 + 6 \cdot 9 + 9 \cdot 12 + \cdots + 3n(3n + 3) = 3n(n + 1)(n + 2)$
9. $2 + 6 + 12 + \cdots + n(n + 1) = \dfrac{n(n + 1)(n + 2)}{3}$
10. $3 + 3^2 + 3^3 + \cdots + 3^n = \dfrac{3(3^n - 1)}{2}$
11. $5 + 10 + 15 + \cdots + 5n = \dfrac{5n(n + 1)}{2}$
12. $a + 2a + 3a + \cdots + na = \dfrac{an(n + 1)}{2}$

appendix

C
TABLES

Number	Square	Square root	Number	Square	Square root
1	1	1.000	11	1 21	3.317
2	4	1.414	12	1 44	3.464
3	9	1.732	13	1 69	3.606
4	16	2.000	14	1 96	3.742
5	25	2.236	15	2 25	3.873
6	36	2.449	16	2 56	4.000
7	49	2.646	17	2 89	4.123
8	64	2.828	18	3 24	4.243
9	81	3.000	19	3 61	4.359
10	1 00	3.162	20	4 00	4.472

TABLE OF SQUARES AND SQUARE ROOTS (*Continued*)

Number	Square	Square root	Number	Square	Square root
21	4 41	4.583	56	31 36	7.483
22	4 84	4.690	57	32 49	7.550
23	5 29	4.796	58	33 64	7.616
24	5 76	4.899	59	34 81	7.681
25	6 25	5.000	60	36 00	7.746
26	6 76	5.099	61	37 21	7.810
27	7 29	5.196	62	38 44	7.874
28	7 84	5.292	63	39 69	7.937
29	8 41	5.385	64	40 96	8.000
30	9 00	5.477	65	42 25	8.062
31	9 61	5.568	66	43 56	8.124
32	10 24	5.657	67	44 89	8.185
33	10 89	5.745	68	46 24	8.246
34	11 56	5.831	69	47 61	8.307
35	12 25	5.916	70	49 00	8.367
36	12 96	6.000	71	50 41	8.426
37	13 69	6.083	72	51 84	8.485
38	14 44	6.164	73	53 29	8.544
39	15 21	6.245	74	54 76	8.602
40	16 00	6.325	75	56 25	8.660
41	16 81	6.403	76	57 76	8.718
42	17 64	6.481	77	59 29	8.775
43	18 49	6.557	78	60 84	8.832
44	19 36	6.633	79	62 41	8.888
45	20 25	6.708	80	64 00	8.944
46	21 16	6.782	81	65 61	9.000
47	22 09	6.856	82	67 24	9.055
48	23 04	6.928	83	68 89	9.110
49	24 01	7.000	84	70 56	9.165
50	25 00	7.071	85	72 25	9.220
51	26 01	7.141	86	73 96	9.274
52	27 04	7.211	87	75 69	9.327
53	28 09	7.280	88	77 44	9.381
54	29 16	7.348	89	79 21	9.434
55	30 25	7.416	90	81 00	9.487

TABLE OF SQUARES AND SQUARE ROOTS (*Continued*)

Number	Square	Square root	Number	Square	Square root
91	82 81	9.539	126	1 58 76	11.225
92	84 64	9.592	127	1 61 29	11.269
93	86 49	9.644	128	1 63 84	11.314
94	88 36	9.695	129	1 66 41	11.358
95	90 25	9.747	130	1 69 00	11.402
96	92 16	9.798	131	1 71 61	11.446
97	94 09	9.849	132	1 74 24	11.489
98	96 04	9.899	133	1 76 89	11.533
99	98 01	9.950	134	1 79 56	11.576
100	1 00 00	10.000	135	1 82 25	11.619
101	1 02 01	10.050	136	1 84 96	11.662
102	1 04 04	10.100	137	1 87 69	11.705
103	1 06 09	10.149	138	1 90 44	11.747
104	1 08 16	10.198	139	1 93 21	11.790
105	1 10 25	10.247	140	1 96 00	11.832
106	1 12 36	10.296	141	1 98 81	11.874
107	1 14 49	10.344	142	2 01 64	11.916
108	1 16 64	10.392	143	2 04 49	11.958
109	1 18 81	10.440	144	2 07 36	12.000
110	1 21 00	10.488	145	2 10 25	12.042
111	1 23 21	10.536	146	2 13 16	12.083
112	1 25 44	10.583	147	2 16 09	12.124
113	1 27 69	10.630	148	2 19 04	12.166
114	1 29 96	10.776	149	2 22 01	12.207
115	1 32 25	10.724	150	2 25 00	12.247
116	1 34 56	10.770	151	2 28 01	12.288
117	1 36 89	10.817	152	2 31 04	12.329
118	1 39 24	10.863	153	2 34 09	12.369
119	1 41 61	10.909	154	2 37 16	12.410
120	1 44 00	10.954	155	2 40 25	12.450
121	1 46 41	11.000	156	2 43 36	12.490
122	1 48 84	11.045	157	2 46 49	12.530
123	1 51 29	11.091	158	2 49 64	12.570
124	1 53 76	11.136	159	2 52 81	12.610
125	1 56 25	11.180	160	2 56 00	12 649

TABLE OF SQUARES AND SQUARE ROOTS (*Continued*)

Number	Square	Square root	Number	Square	Square root
161	2 59 21	12.689	196	3 84 16	14.000
162	2 62 44	12.728	197	3 88 09	14.036
163	2 65 69	12.767	198	3 92 04	14.071
164	2 68 96	12.806	199	3 96 01	14.107
165	2 72 25	12.845	200	4 00 00	14.142
166	2 75 56	12.884	201	4 04 01	14.177
167	2 78 89	12.923	202	4 08 04	14.213
168	2 82 24	12.961	203	4 12 09	14.248
169	2 85 61	13.000	204	4 16 16	14.283
170	2 89 00	13.038	205	4 20 25	14.318
171	2 92 41	13.077	206	4 24 36	14.353
172	2 95 84	13.115	207	4 28 49	14.387
173	2 99 29	13.153	208	4 32 64	14.422
174	3 02 76	13.191	209	4 36 81	14.457
175	3 06 25	13.229	210	4 41 00	14.491
176	3 09 76	13.266	211	4 45 21	14.526
177	3 13 29	13.304	212	4 49 44	14.560
178	3 16 84	13.342	213	4 53 69	14.595
179	3 20 41	13.379	214	4 57 96	14.629
180	3 24 00	13.416	215	4 62 25	14.663
181	3 27 61	13.454	216	4 66 56	14.697
182	3 31 24	13.491	217	4 70 89	14.731
183	3 34 89	13.528	218	4 75 24	14.765
184	3 38 56	13.565	219	4 79 61	14.799
185	3 42 25	13.601	220	4 84 00	14.832
186	3 45 96	13.638	221	4 88 41	14.866
187	3 49 69	13.675	222	4 92 84	14.900
188	3 53 44	13.711	223	4 97 29	14.933
189	3 57 21	13.748	224	5 01 76	14.967
190	3 61 00	13.784	225	5 06 25	15.000
191	3 64 81	13.820	226	5 10 76	15.033
192	3 68 64	13.856	227	5 15 29	15.067
193	3 72 49	13.892	228	5 19 84	15.100
194	3 76 36	13.928	229	5 24 41	15.133
195	3 80 25	13.964	230	5 29 00	15.166

TABLE OF SQUARES AND SQUARE ROOTS (*Continued*)

Number	Square	Square root	Number	Square	Square root
231	5 33 61	15.199	266	7 07 56	16.310
232	5 38 24	15.232	267	7 12 89	16.340
233	5 42 89	15.264	268	7 18 24	16.371
234	5 47 56	15.297	269	7 23 61	16.401
235	5 52 25	15.330	270	7 29 00	16.432
236	5 56 96	15.362	271	7 34 41	16.462
237	5 61 69	15.395	272	7 39 84	16.492
238	5 66 44	15.427	273	7 45 29	16.523
239	5 71 21	15.460	274	7 50 76	16.553
240	5 76 00	15.492	275	7 56 25	16.583
241	5 80 81	15.524	276	7 61 76	16.613
242	5 85 64	15.556	277	7 67 29	16.643
243	5 90 49	15.588	278	7 72 84	16.673
244	5 95 36	15.620	279	7 78 41	16.703
245	6 00 25	15.652	280	7 84 00	16.733
246	6 05 16	15.684	281	7 89 61	16.763
247	6 10 09	15.716	282	7 95 24	16.793
248	6 15 04	15.748	283	8 00 89	16.823
249	6 20 01	15.780	284	8 06 56	16.852
250	6 25 00	15.811	285	8 12 25	16.882
251	6 30 01	15.843	286	8 17 96	16.912
252	6 35 04	15.875	287	8 23 69	16.941
253	6 40 09	15.906	288	8 29 44	16.971
254	6 45 16	15.937	289	8 35 21	17.000
255	6 50 25	15.969	290	8 41 00	17.029
256	6 55 36	16.000	291	8 46 81	17.059
257	6 60 49	16.031	292	8 52 64	17.088
258	6 65 64	16.062	293	8 58 49	17.117
259	6 70 81	16.093	294	8 64 36	17.146
260	6 76 00	16.125	295	8 70 25	17.176
261	6 81 21	16.155	296	8 76 16	17.205
262	6 86 44	16.186	297	8 82 09	17.234
263	6 91 69	16.217	298	8 88 04	17.263
264	6 96 96	16.248	299	8 94 01	17.292
265	7 02 25	16.279	300	9 00 00	17.321

TABLE OF SQUARES AND SQUARE ROOTS (*Continued*)

Number	Square	Square root	Number	Square	Square root
301	9 06 01	17.349	336	11 28 96	18.330
302	9 12 04	17.378	337	11 35 69	18.358
303	9 18 09	17.407	338	11 42 44	18.385
304	9 24 16	17.436	339	11 49 21	18.412
305	9 30 25	17.464	340	11 56 00	18.439
306	9 36 36	17.493	341	11 62 81	18.466
307	9 42 49	17.521	342	11 69 64	18.493
308	9 48 64	17.550	343	11 76 49	18.520
309	9 54 81	17.578	344	11 83 36	18.547
310	9 61 00	17.607	345	11 90 25	18.574
311	9 67 21	17.635	346	11 97 16	18.601
312	9 73 44	17.664	347	12 04 09	18.628
313	9 79 69	17.692	348	12 11 04	18.655
314	9 85 96	17.720	349	12 18 01	18.682
315	9 92 25	17.748	350	12 25 00	18.708
316	9 98 56	17.776	351	12 32 01	18.735
317	10 04 89	17.804	352	12 39 04	18.762
318	10 11 24	17.833	353	12 46 09	18.788
319	10 17 61	17.861	354	12 53 16	18.815
320	10 24 00	17.889	355	12 60 25	18.841
321	10 30 41	17.916	356	12 67 36	18.868
322	10 36 84	17.944	357	12 74 49	18.894
323	10 43 29	17.972	358	12 81 64	18.921
324	10 49 76	18.000	359	12 88 81	18.947
325	10 56 25	18.028	360	12 96 00	18.974
326	10 62 76	18.055	361	13 03 21	19.000
327	10 69 29	18.083	362	13 10 44	19.026
328	10 75 84	18.111	363	13 17 69	19.053
329	10 82 41	18.138	364	13 24 96	19.079
330	10 89 00	18.166	365	13 32 25	19.105
331	10 95 61	18.193	366	13 39 56	19.131
332	11 02 24	18.221	367	13 46 89	19.157
333	11 08 89	18.248	368	13 54 24	19.183
334	11 15 56	18.276	369	13 61 61	19.209
335	11 22 25	18.303	370	13 69 00	19.235

TABLE OF SQUARES AND SQUARE ROOTS (*Continued*)

Number	Square	Square root	Number	Square	Square root
371	13 76 41	19.261	406	16 48 36	20.149
372	13 83 84	19.287	407	16 56 49	20.174
373	13 91 29	19.313	408	16 64 64	20.199
374	13 98 76	19.339	409	16 72 81	20.224
375	14 06 25	19.363	410	16 81 00	20.248
376	14 13 76	19.391	411	16 89 21	20.273
377	14 21 29	19.416	412	16 97 44	20.298
378	14 28 84	19.442	413	17 05 69	20.322
379	14 36 41	19.468	414	17 13 96	20.347
380	14 44 00	19.494	415	17 22 25	20.372
381	14 51 61	19.519	416	17 30 56	20.396
382	14 59 24	19.545	417	17 38 89	20.421
383	14 66 89	19.570	418	17 47 24	20.445
384	14 74 56	19.596	419	17 55 61	20.469
385	14 82 25	19.621	420	17 64 00	20.494
386	14 89 96	19.647	421	17 72 41	20.518
387	14 97 69	19.672	422	17 80 84	20.543
388	15 05 44	19.698	423	17 89 29	20.567
389	15 13 21	19.723	424	17 97 76	20.591
390	15 21 00	19.748	425	18 06 25	20.616
391	15 28 81	19.774	426	18 14 76	20.640
392	15 36 64	19.799	427	18 23 29	20.664
393	15 44 49	19.824	428	18 31 84	20.688
394	15 52 36	19.849	429	18 40 41	20 712
395	15 60 25	19.875	430	18 49 00	20.736
396	15 68 16	19.900	431	18 57 61	20.761
397	15 76 09	19.925	432	18 66 24	20.785
398	15 84 04	19.950	433	18 74 89	20.809
399	15 92 01	19.975	434	18 83 56	20.833
400	16 00 00	20.000	435	18 92 25	20.857
401	16 08 01	20.025	436	19 00 96	20.881
402	16 16 04	20.050	437	19 09 69	20.905
403	16 24 09	20.075	438	19 18 44	20.928
404	16 32 16	20.100	439	19 27 21	20.952
405	16 40 25	20.125	440	19 36 00	20.976

TABLE OF SQUARES AND SQUARE ROOTS (*Continued*)

Number	Square	Square root	Number	Square	Square root
441	19 44 81	21.000	476	22 65 76	21.817
442	19 53 64	21.024	477	22 75 29	21.840
443	19 62 49	21.048	478	22 84 84	21.863
444	19 71 36	21.071	479	22 94 41	21.886
445	19 80 25	21.095	480	23 04 00	21.909
446	19 89 16	21.119	481	23 13 61	21.932
447	19 98 09	21.142	482	23 23 24	21.954
448	20 07 04	21.166	483	23 32 89	21.977
449	20 16 01	21.190	484	23 42 56	22.000
450	20 25 00	21.213	485	23 52 25	22.023
451	20 34 01	21.237	486	23 61 96	22.045
452	20 43 04	21.260	487	23 71 69	22.068
453	20 52 09	21.284	488	23 81 44	22.091
454	20 61 16	21.307	489	23 91 21	22.113
455	20 70 25	21.831	490	24 01 00	22.136
456	20 79 36	21.354	491	24 10 81	22.159
457	20 88 49	21.378	492	24 20 64	22.181
458	20 97 64	21.401	493	24 30 49	22.204
459	21 06 81	21.424	494	24 40 36	22.226
460	21 16 00	21.448	495	24 50 25	22.249
461	21 25 21	21.471	496	24 60 16	22.271
462	21 34 44	21.494	497	24 70 09	22.293
463	21 43 69	21.517	498	24 80 04	22.316
464	21 52 96	21.541	499	24 90 01	22.338
465	21 62 25	21.564	500	25 00 00	22.361
466	21 71 56	21.587	501	25 10 01	22.383
467	21 80 89	21.610	502	25 20 04	22.405
468	21 90 24	21.633	503	25 30 09	22.428
469	21 99 61	21.656	504	25 40 16	22.450
470	22 09 00	21.679	505	25 50 25	22.472
471	22 18 41	21.703	506	25 60 36	22.494
472	22 27 84	21.726	507	25 70 49	22.517
473	22 37 29	21.749	508	25 80 64	22.539
474	22 46 76	21.772	509	25 90 81	22.561
475	22 56 25	21.794	510	26 01 00	22.583

TABLE OF SQUARES AND SQUARE ROOTS (*Continued*)

Number	Square	Square root	Number	Square	Square root
511	26 11 21	22.605	546	29 81 16	23.367
512	26 21 44	22.627	547	29 92 09	23.388
513	26 31 69	22.650	548	30 03 04	23.409
514	26 41 96	22.672	549	30 14 01	23.431
515	26 52 25	22.694	550	30 25 00	23.452
516	26 62 56	22.716	551	30 36 01	23.473
517	26 72 89	22.738	552	30 47 04	23.495
518	26 83 24	22.760	553	30 58 09	23.516
519	26 93 61	22.782	554	30 69 16	23.537
520	27 04 00	22.804	555	30 08 25	23.558
521	27 14 41	22.825	556	30 91 36	23.580
522	27 24 84	22.847	557	31 02 49	23.601
523	27 35 29	22.869	558	31 13 64	23.622
524	27 45 76	22.891	559	31 24 81	23.643
525	27 56 25	22.913	560	31 36 00	23.664
526	27 66 76	22.935	561	31 47 21	23.685
527	27 77 29	22.956	562	31 58 44	23.707
528	27 87 84	22.978	563	31 69 69	23.728
529	27 98 41	23.000	564	31 80 96	23.749
530	28 09 00	23.022	565	31 92 25	23.770
531	28 19 61	23.043	566	32 03 56	23.791
532	28 30 24	23.065	567	32 14 89	23.812
533	28 40 89	23.087	568	32 26 24	23.833
534	28 51 56	23.108	569	32 37 61	23.854
535	28 62 25	23.130	570	32 49 00	23.875
536	28 72 96	23.152	571	32 60 41	23.896
537	28 83 69	23.173	572	32 71 84	23.917
538	28 94 44	23.195	573	32 83 29	23.937
539	29 05 21	23.216	574	32 94 76	23.958
540	29 16 00	23.238	575	33 06 25	23.979
541	29 26 81	23.259	576	33 17 76	24.000
542	29 37 64	23.281	577	33 29 29	24.021
543	29 48 49	23.302	578	33 40 84	24.042
544	29 59 36	23.324	579	33 52 41	24.062
545	29 70 25	23.345	580	33 64 00	24.083

TABLE OF SQUARES AND SQUARE ROOTS (*Continued*)

Number	Square	Square root	Number	Square	Square root
581	33 75 61	24.104	616	37 94 56	24.819
582	33 87 24	24.125	617	38 06 89	24.839
583	33 98 89	24.145	618	38 19 24	24.860
584	34 10 56	24.166	619	38 31 61	24.880
585	34 22 25	24.187	620	38 44 00	24.900
586	34 33 96	24.207	621	38 56 41	24.920
587	34 45 69	24.228	622	38 68 84	24.940
588	34 57 44	24.249	623	38 81 29	24.960
589	34 69 21	24.269	624	38 93 76	24.980
590	34 81 00	24.290	625	39 06 25	25.000
591	34 92 81	24.310	626	39 18 76	25.020
592	35 04 64	24.331	627	39 31 29	25.040
593	35 16 49	24.352	628	39 43 84	25.060
594	35 28 36	24.372	629	39 56 41	25.080
595	35 40 25	24.393	630	39 69 00	25.100
596	35 52 16	24.413	631	39 81 61	25.120
597	35 64 09	24.434	632	39 94 24	25.140
598	35 76 04	24.454	633	40 06 89	25.159
599	35 88 01	24.474	634	40 19 56	25.179
600	36 00 00	24.495	635	40 32 25	25.199
601	36 12 01	24.515	636	40 44 96	25.219
602	36 24 04	24.536	637	40 57 69	25.239
603	36 36 09	24.556	638	40 70 44	25.259
604	36 48 16	24.576	639	40 83 21	25.278
605	36 60 25	24.597	640	40 96 00	25.298
606	36 72 36	24.617	641	41 08 81	25.318
607	36 84 49	24.637	642	41 21 64	25.338
608	36 96 64	24.658	643	41 34 49	25.357
609	37 08 81	24.678	644	41 47 36	25.377
610	37 21 00	24.698	645	41 60 25	25.397
611	37 33 21	24.718	646	41 73 16	25.417
612	37 45 44	24.739	647	41 86 09	25.436
613	37 57 69	24.759	648	41 99 04	25.456
614	37 69 96	24.779	649	42 12 01	25.475
615	37 82 25	24.799	650	42 25 00	25.495

TABLE OF SQUARES AND SQUARE ROOTS (*Continued*)

Number	Square	Square root	Number	Square	Square root
651	42 38 01	25.515	686	47 05 96	26.192
652	42 51 04	25.534	687	47 19 69	26.211
653	42 64 09	25.554	688	47 33 44	26.230
654	42 77 16	25.573	689	47 47 21	26.249
655	42 90 25	25.593	690	47 61 00	26.268
656	43 03 36	25.612	691	47 74 81	26.287
657	43 16 49	25.632	692	47 88 64	26.306
658	43 29 64	25.652	693	48 02 49	26.325
659	43 42 81	25.671	694	48 16 36	26.344
660	43 56 00	25.690	695	48 30 25	26.363
661	43 69 21	25.710	696	48 44 16	26.382
662	43 82 44	25.729	697	48 58 09	26.401
663	43 95 69	25.749	698	48 72 04	26.420
664	44 08 96	25.768	699	48 86 01	26.439
665	44 22 25	25.788	700	49 00 00	26.458
666	44 35 56	25.807	701	49 14 01	26.476
667	44 48 89	25.826	702	49 28 04	26.495
668	44 62 24	25.846	703	49 42 09	26.514
669	44 75 61	25.865	704	49 56 16	26.533
670	44 89 00	25.884	705	49 70 25	26.552
671	45 02 41	25.904	706	49 84 36	26.571
672	45 15 84	25.923	707	49 98 49	26.589
673	45 29 29	25.942	708	50 12 64	26.608
674	45 42 76	25.962	709	50 26 81	26.627
675	45 56 25	25.981	710	50 41 00	26.646
676	45 69 76	26.000	711	50 55 21	26.665
677	45 83 29	26.019	712	50 69 44	26.683
678	45 96 84	26.038	713	50 83 69	26.702
679	46 10 41	26.058	714	50 97 96	26.721
680	46 24 00	26.077	715	51 12 25	26.739
681	46 37 61	26.096	716	51 26 56	26.758
682	46 51 24	26.115	717	51 40 89	26.777
683	46 64 89	26.134	718	51 55 24	26.796
684	46 78 56	26.153	719	51 69 61	26.814
685	46 92 25	26.173	720	51 84 00	26.833

TABLE OF SQUARES AND SQUARE ROOTS (*Continued*)

Number	Square	Square root	Number	Square	Square root
721	51 98 41	26.851	756	57 15 36	27.495
722	52 12 84	26.870	757	57 30 49	27.514
723	52 27 29	26.889	758	57 45 64	27.532
724	52 41 76	26.907	759	57 60 81	27.550
725	52 56 25	26.926	760	57 76 00	27.568
726	52 70 76	26.944	761	57 91 21	27.586
727	52 85 29	26.963	762	58 06 44	27.604
728	52 99 84	26.981	763	58 21 69	27.622
729	53 14 41	27.000	764	58 36 96	27.641
730	53 29 00	27.019	765	58 52 25	27.659
731	53 43 61	27.037	766	58 67 56	27.677
732	53 58 24	27.055	767	58 82 89	27.695
733	53 72 89	27.074	768	58 98 24	27.713
734	53 87 56	27.092	769	59 13 61	27.731
735	54 02 25	27.111	770	59 29 00	27.749
736	54 16 96	27.129	771	59 44 41	27.767
737	54 31 69	27.148	772	59 59 84	27.785
738	54 46 44	27.166	773	59 75 29	27.803
739	54 61 21	27.185	774	59 90 76	27.821
740	54 76 00	27.203	775	60 06 25	27.839
741	54 90 81	27.221	776	60 21 76	27.857
742	55 05 64	27.240	777	60 37 29	27.875
743	55 20 49	27.258	778	60 52 84	27.893
744	55 35 36	27.276	779	60 68 41	27.911
745	55 50 25	27.295	780	60 84 00	27.928
746	55 65 16	27.313	781	60 99 61	27.946
747	55 80 09	27.331	782	61 15 24	27.964
748	55 95 04	27.350	783	61 30 89	27.982
749	56 10 01	27.368	784	61 46 56	28.000
750	56 25 00	27.386	785	61 62 25	28.018
751	56 40 01	27.404	786	61 77 96	28.036
752	56 55 04	27.423	787	61 93 69	28.054
753	56 70 09	27.441	788	62 09 44	28.071
754	56 85 16	27.459	789	62 25 21	28.089
755	57 00 25	27.477	790	62 41 00	28.107

TABLE OF SQUARES AND SQUARE ROOTS (*Continued*)

Number	Square	Square root	Number	Square	Square root
791	62 56 81	28.125	826	68 22 76	28.740
792	62 72 64	28.142	827	68 39 29	28.758
793	62 88 49	28.160	828	68 55 84	28.775
794	63 04 36	28.178	829	68 72 41	28.792
795	63 20 25	28.196	830	68 89 00	28.810
796	63 36 16	28.213	831	69 05 61	28.827
797	63 52 09	28.231	832	69 22 24	28.844
798	63 68 04	28.249	833	69 38 89	28.862
799	63 84 01	28.267	834	69 55 56	28.879
800	64 00 00	28.284	835	69 72 25	28.896
801	64 16 01	28.302	836	69 88 96	28.914
802	64 32 04	28.320	837	70 05 69	28.931
803	64 48 09	28.337	838	70 22 44	28.948
804	64 64 16	28.355	839	70 39 21	28.965
805	64 80 25	28.373	840	70 56 00	28.983
806	64 96 36	28.390	841	70 72 81	29.000
807	65 12 49	28.408	842	70 89 64	29.017
808	65 28 64	28.425	843	71 06 49	29.034
809	65 44 81	28.443	844	71 23 36	29.052
810	65 61 00	28.460	845	71 40 25	29.069
811	65 77 21	28.478	846	71 57 16	29.086
812	65 93 44	28.496	847	71 74 09	29.103
813	66 09 69	28.513	848	71 91 04	29.120
814	66 25 96	28.531	849	72 08 01	29.138
815	66 42 25	28.548	850	72 25 00	29.155
816	66 58 56	28.566	851	72 42 01	29.172
817	66 74 89	28.583	852	72 59 04	29.189
818	66 91 24	28.601	853	72 76 09	29.206
819	67 07 61	28.618	854	72 93 16	29.223
820	67 24 00	28.636	855	73 10 25	29.240
821	67 40 41	28.653	856	73 27 36	29.257
822	67 56 84	28.671	857	73 44 49	29.275
823	67 73 29	28.688	858	73 61 64	29.292
824	67 89 76	28.705	859	73 78 81	29.309
825	68 06 25	28.723	860	73 96 00	29.326

TABLE OF SQUARES AND SQUARE ROOTS (*Continued*)

Number	Square	Square root	Number	Square	Square root
861	74 13 21	29.343	896	80 28 16	29.933
862	74 30 44	29.360	897	80 46 09	29.950
863	74 47 69	29.377	898	80 64 04	29.967
864	74 64 96	29.394	899	80 82 01	29.983
865	74 82 25	29.411	900	81 00 00	30.000
866	74 99 56	29.428	901	81 18 01	30.017
867	75 16 89	29.445	902	81 36 04	30.033
868	75 34 24	29.462	903	81 54 09	30.050
869	75 51 61	29.479	904	81 72 16	30.067
870	75 69 00	29.496	905	81 90 25	30.083
871	75 86 41	29.513	906	82 08 36	30.100
872	76 03 84	29.530	907	82 26 49	30.116
873	76 21 29	29.547	908	82 44 64	30.133
874	76 38 76	29.563	909	82 62 81	30.150
875	76 56 25	29.580	910	82 81 00	30.166
876	76 73 76	29.597	911	82 99 21	30.183
877	76 91 29	29.614	912	83 17 44	30.199
878	77 08 84	29.631	913	83 35 69	30.216
879	77 26 41	29.648	914	83 53 96	30.232
880	77 44 00	29.665	915	83 72 25	30.249
881	77 61 61	29.682	916	83 90 56	30.265
882	77 79 24	29.698	917	84 08 89	30.282
883	77 96 89	29.715	918	84 27 24	30.299
884	78 14 56	29.732	919	84 45 61	30.315
885	78 32 25	29.749	920	84 64 00	30.332
886	78 49 96	29.766	921	84 82 41	30.348
887	78 67 69	29.783	922	85 00 84	30.364
888	78 85 44	29.799	923	85 19 29	30.381
889	79 03 21	29.816	924	85 37 76	30.397
890	79 21 00	29.833	925	85 56 25	30.414
891	79 38 81	29.850	926	85 74 76	30.430
892	79 56 64	29.866	927	85 93 29	30.447
893	79 74 49	29.883	928	86 11 84	30.463
894	79 92 36	29.900	929	86 30 41	30.480
895	80 10 25	29.916	930	86 49 00	30.496

TABLE OF SQUARES AND SQUARE ROOTS (*Continued*)

Number	Square	Square root	Number	Square	Square root
931	86 67 61	30.512	966	93 31 50	31.081
932	86 86 24	30.529	967	93 50 89	31.097
933	87 04 89	30.545	968	93 70 24	31.113
934	87 23 56	30.561	969	93 89 61	31.129
935	87 42 25	30.578	970	94 09 00	31.145
936	87 60 96	30.594	971	94 28 41	31.161
937	87 79 69	30.610	972	94 47 84	31.177
938	87 98 44	30.627	973	94 67 29	31.193
939	88 17 21	30.643	974	94 86 76	31.209
940	88 36 00	30.659	975	95 06 25	31.225
941	88 54 81	30.676	976	95 25 76	31.241
942	88 73 64	30.692	977	95 45 29	31.257
943	88 92 49	30.708	978	95 64 84	31.273
944	89 11 36	30.725	979	95 84 41	31.289
945	89 30 25	30.741	980	96 04 00	31.305
946	89 49 16	30.757	981	96 23 61	31.321
947	89 68 09	30.773	982	96 43 24	31.337
948	89 87 04	30.790	983	96 62 89	31.353
949	90 06 01	30.806	984	96 82 56	31.369
950	90 25 00	30.822	985	97 02 25	31.385
951	90 44 01	30.838	986	97 21 96	31.401
952	90 63 04	30.854	987	97 41 69	31.417
953	90 82 09	30.871	988	97 61 44	31.432
954	91 01 16	30 887	989	97 81 21	31.448
955	91 20 25	30.903	990	98 01 00	31.464
956	91 39 36	30.919	991	98 20 81	31.480
957	91 58 49	30.935	992	98 40 64	31.496
958	91 77 64	30.952	993	98 60 49	31.512
959	91 96 81	30.968	994	98 80 36	31.528
960	92 16 00	30.984	995	99 00 25	31.544
961	92 35 21	31.000	996	99 20 16	31.559
962	92 54 44	31.016	997	99 40 09	31.575
963	92 73 69	31.032	998	99 60 04	31.591
964	92 92 96	31.048	999	99 80 01	31.607
965	93 12 25	31.064	1000	100 00 00	31.623

TABLE C.2 AREAS UNDER THE NORMAL CURVE*

z	0.00	0.01	0.02	0.03	0.04	0.05	0.06	0.07	0.08	0.09
-3.4	0.0003	0.0003	0.0003	0.0003	0.0003	0.0003	0.0003	0.0003	0.0003	0.0002
-3.3	0.0005	0.0005	0.0005	0.0004	0.0004	0.0004	0.0004	0.0004	0.0004	0.0003
-3.2	0.0007	0.0007	0.0006	0.0006	0.0006	0.0006	0.0006	0.0005	0.0005	0.0005
-3.1	0.0010	0.0009	0.0009	0.0009	0.0008	0.0008	0.0008	0.0008	0.0007	0.0007
-3.0	0.0013	0.0013	0.0013	0.0012	0.0012	0.0011	0.0011	0.0011	0.0010	0.0010
-2.9	0.0019	0.0018	0.0017	0.0017	0.0016	0.0016	0.0015	0.0015	0.0014	0.0014
-2.8	0.0026	0.0025	0.0024	0.0023	0.0023	0.0022	0.0021	0.0021	0.0020	0.0019
-2.7	0.0035	0.0034	0.0033	0.0032	0.0031	0.0030	0.0029	0.0028	0.0027	0.0026
-2.6	0.0047	0.0045	0.0044	0.0043	0.0041	0.0040	0.0039	0.0038	0.0037	0.0036
-2.5	0.0062	0.0060	0.0059	0.0057	0.0055	0.0054	0.0052	0.0051	0.0049	0.0048
-2.4	0.0082	0.0080	0.0078	0.0075	0.0073	0.0071	0.0069	0.0068	0.0066	0.0064
-2.3	0.0107	0.0104	0.0102	0.0099	0.0096	0.0094	0.0091	0.0089	0.0087	0.0084
-2.2	0.0139	0.0136	0.0132	0.0129	0.0125	0.0122	0.0119	0.0116	0.0113	0.0110
-2.1	0.0179	0.0174	0.0170	0.0166	0.0162	0.0158	0.0154	0.0150	0.0146	0.0143
-2.0	0.0228	0.0222	0.0217	0.0212	0.0207	0.0202	0.0197	0.0192	0.0188	0.0183
-1.9	0.0287	0.0281	0.0274	0.0268	0.0262	0.0256	0.0250	0.0244	0.0239	0.0233
-1.8	0.0359	0.0352	0.0344	0.0336	0.0329	0.0322	0.0314	0.0307	0.0301	0.0294
-1.7	0.0446	0.0436	0.0427	0.0418	0.0409	0.0401	0.0392	0.0384	0.0375	0.0367
-1.6	0.0548	0.0537	0.0526	0.0516	0.0505	0.0495	0.0485	0.0475	0.0465	0.0455
-1.5	0.0668	0.0655	0.0643	0.0630	0.0618	0.0606	0.0594	0.0582	0.0571	0.0559
-1.4	0.0808	0.0793	0.0778	0.0764	0.0749	0.0735	0.0722	0.0708	0.0694	0.0681
-1.3	0.0968	0.0951	0.0934	0.0918	0.0901	0.0885	0.0869	0.0853	0.0838	0.0823
-1.2	0.1151	0.1131	0.1112	0.1093	0.1075	0.1056	0.1038	0.1020	0.1003	0.0985
-1.1	0.1357	0.1335	0.1314	0.1292	0.1271	0.1251	0.1230	0.1210	0.1190	0.1170
-1.0	0.1587	0.1562	0.1539	0.1515	0.1492	0.1469	0.1446	0.1423	0.1401	0.1379
-0.9	0.1841	0.1814	0.1788	0.1762	0.1736	0.1711	0.1685	0.1660	0.1635	0.1611
-0.8	0.2119	0.2090	0.2061	0.2033	0.2005	0.1977	0.1949	0.1922	0.1894	0.1867
-0.7	0.2420	0.2389	0.2358	0.2327	0.2296	0.2266	0.2236	0.2206	0.2177	0.2148
-0.6	0.2743	0.2709	0.2676	0.2643	0.2611	0.2578	0.2546	0.2514	0.2483	0.2451
-0.5	0.3085	0.3050	0.3015	0.2981	0.2946	0.2912	0.2877	0.2843	0.2810	0.2776
-0.4	0.3446	0.3409	0.3372	0.3336	0.3300	0.3264	0.3228	0.3192	0.3156	0.3121
-0.3	0.3821	0.3783	0.3745	0.3707	0.3669	0.3632	0.3594	0.3557	0.3520	0.3483
-0.2	0.4207	0.4168	0.4129	0.4090	0.4052	0.4013	0.3974	0.3936	0.3897	0.3859
-0.1	0.4602	0.4562	0.4522	0.4483	0.4443	0.4404	0.4364	0.4325	0.4286	0.4247
-0.0	0.5000	0.4960	0.4920	0.4880	0.4840	0.4801	0.4761	0.4721	0.4681	0.4641
0.0	0.5000	0.5040	0.5080	0.5120	0.5160	0.5199	0.5239	0.5279	0.5319	0.5359
0.1	0.5398	0.5438	0.5478	0.5517	0.5557	0.5596	0.5636	0.5675	0.5714	0.5753
0.2	0.5793	0.5832	0.5871	0.5910	0.5948	0.5987	0.6026	0.6064	0.6103	0.6141
0.3	0.6179	0.6217	0.6244	0.6293	0.6331	0.6368	0.6406	0.6443	0.6480	0.6517
0.4	0.6554	0.6591	0.6628	0.6664	0.6700	0.6736	0.6772	0.6808	0.6844	0.6879
0.5	0.6915	0.6950	0.6985	0.7019	0.7054	0.7088	0.7123	0.7157	0.7190	0.7224
0.6	0.7257	0.7291	0.7324	0.7357	0.7389	0.7422	0.7454	0.7486	0.7517	0.7549
0.7	0.7580	0.7611	0.7642	0.7673	0.7704	0.7734	0.7764	0.7794	0.7823	0.7852
0.8	0.7881	0.7910	0.7939	0.7967	0.7995	0.8023	0.8051	0.8078	0.8106	0.8133
0.9	0.8159	0.8186	0.8212	0.8238	0.8264	0.8289	0.8315	0.8340	0.8365	0.8389
1.0	0.8413	0.8438	0.8461	0.8485	0.8508	0.8531	0.8554	0.8577	0.8599	0.8621
1.1	0.8643	0.8665	0.8686	0.8708	0.8729	0.8749	0.8770	0.8790	0.8810	0.8830
1.2	0.8849	0.8869	0.8888	0.8907	0.8925	0.8944	0.8962	0.8980	0.8997	0.9015
1.3	0.9032	0.9049	0.9066	0.9082	0.9099	0.9115	0.9131	0.9147	0.9162	0.9177
1.4	0.9192	0.9207	0.9222	0.9236	0.9251	0.9265	0.9278	0.9292	0.9306	0.9319
1.5	0.9332	0.9345	0.9357	0.9370	0.9382	0.9394	0.9406	0.9418	0.9429	0.9441
1.6	0.9452	0.9463	0.9474	0.9484	0.9495	0.9505	0.9515	0.9525	0.9535	0.9545
1.7	0.9554	0.9564	0.9573	0.9582	0.9591	0.9599	0.9608	0.9616	0.9625	0.9633
1.8	0.9641	0.9649	0.9656	0.9664	0.9671	0.9678	0.9686	0.9693	0.9699	0.9706
1.9	0.9713	0.9719	0.9726	0.9732	0.9738	0.9744	0.9750	0.9756	0.9761	0.9767
2.0	0.9772	0.9778	0.9783	0.9788	0.9793	0.9798	0.9803	0.9808	0.9812	0.9817
2.1	0.9821	0.9826	0.9830	0.9834	0.9838	0.9842	0.9846	0.9850	0.9854	0.9857
2.2	0.9861	0.9864	0.9868	0.9871	0.9875	0.9878	0.9881	0.9884	0.9887	0.9890
2.3	0.9893	0.9896	0.9898	0.9901	0.9904	0.9906	0.9909	0.9911	0.9913	0.9916
2.4	0.9918	0.9920	0.9922	0.9925	0.9927	0.9929	0.9931	0.9932	0.9934	0.9936
2.5	0.9938	0.9940	0.9941	0.9943	0.9945	0.9946	0.9948	0.9949	0.9951	0.9952
2.6	0.9953	0.9955	0.9956	0.9957	0.9959	0.9960	0.9961	0.9962	0.9963	0.9964
2.7	0.9965	0.9966	0.9967	0.9968	0.9969	0.9970	0.9971	0.9972	0.9973	0.9974
2.8	0.9974	0.9975	0.9976	0.9977	0.9977	0.9978	0.9979	0.9979	0.9980	0.9981
2.9	0.9981	0.9982	0.9982	0.9983	0.9984	0.9984	0.9985	0.9985	0.9986	0.9986
3.0	0.9987	0.9987	0.9987	0.9988	0.9988	0.9989	0.9989	0.9989	0.9990	0.9990
3.1	0.9990	0.9991	0.9991	0.9991	0.9992	0.9992	0.9992	0.9992	0.9993	0.9993
3.2	0.9993	0.9993	0.9994	0.9994	0.9994	0.9994	0.9994	0.9995	0.9995	0.9995
3.3	0.9995	0.9995	0.9995	0.9996	0.9996	0.9996	0.9996	0.9996	0.9996	0.9997
3.4	0.9997	0.9997	0.9997	0.9997	0.9997	0.9997	0.9997	0.9997	0.9997	0.9998

* From *Fundamentals of Finite Mathematics* by Volker and Wargo. Copyright 1972, Intext Educational Publishers. Used with permission of Intext Educational Publishers.

appendix

D

SOLUTIONS TO
SELECTED EXERCISES

Exercise Set 1.1

1. (a) Statement
   (c) Not a statement
   (e) Statement
2. (a) I am not broke.
   (c) Dating is not fun.
   (e) Married men do not live longer than single men.
   (g) No politicians are honest.
   (i) Not everyone likes mathematics.
   (k) The Cardinals did not win the game.
3. When P is false

Exercise Set 1.2

1. (a) I will try, and I will succeed.
    (c) I will try, but I will not succeed.
    (e) I will try, or I will not succeed.
    (g) I will not try or I will succeed.
    (i) I will not try, and I will not succeed.
    (k) I will not try, or I will not succeed.
    (m) I will try, and I will not succeed.

2. (a) $p \wedge \sim q$
    (c) $p \wedge q$
    (e) $\sim (\sim p \vee \sim q)$
    (g) $(p \wedge \sim q) \vee (q \wedge \sim p)$
    (i) $\sim (\sim p \wedge \sim q)$

3. (a) False
    (c) True
    (e) True

Exercise Set 1.3

1. (a) $p \rightarrow q$ (g) $p \rightarrow (g \vee h)$
    (c) $(p \vee q) \rightarrow h$ (i) $p \leftrightarrow (q \vee h)$
    (e) $(p \wedge q) \rightarrow h$

2. (a) If I do not like mathematics and I do not like physics, then I
         do not like mathematics.
    (c) If I do not like mathematics or physics, then I do not like
         physics.
    (e) If I like mathematics and I like physics, then I like mathema-
         tics or I like physics.
    (g) If I like mathematics and I like physics, then I like mathema-
         tics.
    (i) If I do not like mathematics, then I do not like mathematics
         and physics.

3. (a) Negation   (g) Inverse
    (c) None   (i) Contrapositive
    (e) None

4. (a) Converse: If you enjoy class, then you do well in it.
         Inverse: If you do not do well in class, then you do not
             enjoy it.
         Contrapositive: If you do not enjoy class, then you do not
             do well in it.

(c) Converse: If we cancel the football game, then it will have rained.

Inverse: If it does not rain, then we shall not cancel the football game.

Contrapositive: If we do not cancel the football game, it will not have rained.

(g) Converse: $\sim q \rightarrow \sim p$ (i) Converse: $\sim p \rightarrow \sim q$

Inverse: $p \rightarrow q$ Inverse: $q \rightarrow p$

Contrapositive: $q \rightarrow p$ Contrapositive: $p \rightarrow q$

Exercise Set 1.4

1. (a)

p	q	$\sim p$	$\sim q$	$\sim p \vee q$
T	T	F	F	T
T	F	F	T	F
F	T	T	F	T
F	F	T	T	T

(c)

p	$\sim p$	$p \vee \sim p$
T	F	T
F	T	T

(e)

p	q	$p \wedge q$	$\sim (p \wedge q)$
T	T	T	F
T	F	F	T
F	T	F	T
F	F	F	T

(g)

p	q	$p \vee q$	$\sim (p \vee q)$
T	T	T	F
T	F	T	F
F	T	T	F
F	F	F	T

(i)

p	q	$p \vee q$	$p \wedge q$	$\sim (p \vee q)$	$\sim (p \wedge q)$	$\sim (p \vee q) \vee \sim (p \wedge q)$ $\rightarrow \sim (p \wedge q)$
T	T	T	T	F	F	T
T	F	T	F	F	T	T
F	T	T	F	F	T	T
F	F	F	F	T	T	T

(k)

p	q	$\sim p$	$\sim q$	$p \rightarrow q$	$\sim q \rightarrow \sim p$	$(p \rightarrow q) \leftrightarrow (\sim q \rightarrow \sim p)$
T	T	F	F	T	T	T
T	F	F	T	F	F	T
F	T	T	F	T	T	T
F	F	T	T	T	T	T

(m)

p	q	$p \wedge q$	$p \vee q$	$\sim(p \wedge q)$	$\sim(p \vee q)$	$\sim(p \wedge q)$ $\to \sim(p \vee q)$
T	T	T	T	F	F	T
T	F	F	T	T	F	F
F	T	F	T	T	F	F
F	F	F	F	T	T	T

(o)

p	q	$\sim p$	$p \vee q$	$(p \vee q) \wedge \sim p$	$(p \vee q) \wedge \sim p \to q$
T	T	F	T	F	T
T	F	F	T	F	T
F	T	T	T	T	T
F	F	T	F	F	T

(q)

p	q	$p \vee q$	$q \to (p \vee q)$
T	T	T	T
T	F	T	T
F	T	T	T
F	F	F	T

(s)

p	q	$(p \to q)$	$(q \to p)$	$(p \to q) \wedge (q \to p)$
T	T	T	T	T
T	F	F	T	F
F	T	T	F	F
F	F	T	T	T

2. (a) No
   (c) Yes
   (e) No
   (g) No
   (i) Yes

   (k) Yes
   (m) No
   (o) Yes
   (q) Yes
   (s) No

3. (a) True
   (c) False
   (e) True
   (g) True
   (i) False

   (k) False
   (m) False
   (o) False
   (q) True
   (s) True

Exercise Set 1.5

1. (a) Valid
   (c) Invalid
   (e) Invalid

   (g) Valid
   (i) Valid

2. (a) Invalid
   (c) Valid
   (e) Valid

   (g) Valid
   (i) Invalid

3. (a) $p \rightarrow q$

    $\underline{p}$

    $\therefore q$

    (c) $p \rightarrow q$

    $\underline{\sim p}$

    $\therefore \sim q$

    (e) $p \rightarrow \sim q$

    $\underline{\sim p}$

    $\therefore q$

    (g) $p \wedge q \rightarrow r$

    $\underline{p \wedge q}$

    $\therefore r$

    (i) $p \rightarrow \sim q$

    $r \rightarrow q$

    $\underline{r}$

    $\therefore \sim p$

4. (a) No farmers are politicians.
    (c) Mr. Vance will buy stock.
    (e) No valid conclusion.
    (g) My uncle is not a forest ranger.
    (i) One who has the strength to stand by his conviction grows through the experience.
    (k) I did not get cut from the bowling team.

Exercise Set 1.6

1. (a)

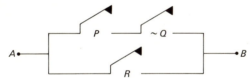

    (c)

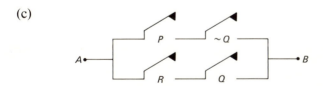

    (e)

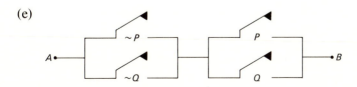

(g)

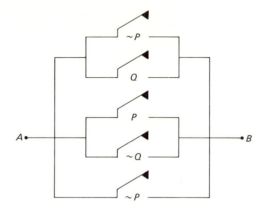

(i)

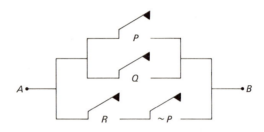

2. (a) $p \wedge (q \vee r)$
    (c) $(p \vee q) \wedge [(r \wedge \sim q) \vee \sim p]$
    (e) $[(p \wedge q) \vee r \vee (\sim p \wedge \sim q)] \wedge \sim r$
3. (a) True
    (c) True
    (e) False
4. (a)

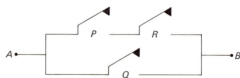

    (c)

(e)

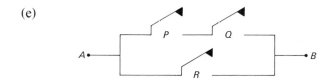

5.

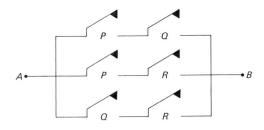

Review Exercise Set 1.7

1. (a) This world will sometime be free of war.
    (c) Some sport is more popular than football in America today.
    (e) All people care very much about something.

2. (a) Converse: If you contribute to air pollution, then you smoke.

    Inverse: If you do not smoke, then you do not contribute to air pollution.

    Contrapositive: If you do not contribute to air pollution, then you do not smoke.

    (c) Converse: If you are lucky, then you enjoy your job.

    Inverse: If you do not enjoy your job, then you are not lucky.

    Contrapositive: If you are not lucky, then you do not enjoy your job.

    (e) Converse: If you are temporarily insane, then you are angry.

    Inverse: If you are not angry, then you are not temporarily insane.

    Contrapositive: If you are not temporarily insane, then you are not angry.

3. (a)

p	q	$\sim q$	$p \wedge \sim q$
T	T	F	F
T	F	T	T
F	T	F	F
F	F	T	F

(c)

p	q	$(p \lor q)$	$(p \lor q) \to p$
T	T	T	T
T	F	T	T
F	T	T	F
F	F	F	T

(e)

p	q	$\sim p$	$\sim q$	$p \land q$	$[p \land q \lor \sim p]$	$[(p \land q) \lor \sim p \to q]$
T	T	F	F	T	T	T
T	F	F	T	F	F	T
F	T	T	F	F	T	T
F	F	T	T	F	T	F

4. (a) Valid

$$p \to q$$
$$\underline{p}$$
$$\therefore q$$

(c) Invalid

(e) Valid

$$p \to q$$
$$q \to r$$
$$r \to s$$
$$\underline{s \to t}$$
$$\therefore p \to t$$

Exercise Set 2.1

1. (a) True   (i) False
   (c) False   (k) False
   (e) False   (m) True
   (g) True   (o) True

2. (a) $\varnothing$, $\{0\}$
   (c) $\varnothing$, $\{a\}$, $\{b\}$, $\{c\}$, $\{a,b\}$, $\{a,c\}$, $\{b,c\}$, $\{a,b,c\}$
   (e) $\varnothing$, $\{\varnothing\}$

3. (a) $\{4,5,6,7\}$ (g) $\{0,1,2,3,6,7,8,9,10\}$
   (c) $\{0,3,6,9\}$ (i) $\{1,2,3,4,5,6,7,8,9,10\}$
   (e) $\{0\}$

5.

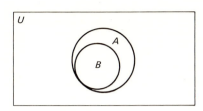

Exercise Set 2.2

1.   (a) {1,2,3,5,7,9}
     (c) {0,2,3,4,5,6,7,8,10}
     (e) {2}
     (g) {2}
     (i) ∅
     (k) {0,1,9}
     (m) {4,6,8,10}
     (o) {4,6,8,10}
     (q) {0,1,2,4,6,8,9,10}
     (s) *U*

2.   (a)   (c)

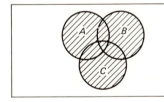   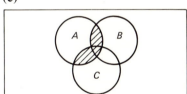

     (e)   (g)

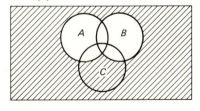   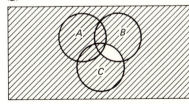

(i)

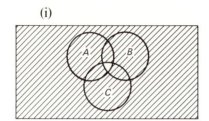

(k)

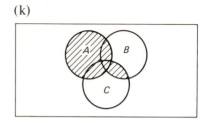

(m)

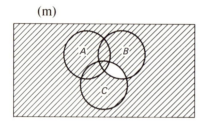

(o)

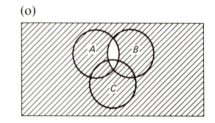

3. (a) False
    (c) True
    (e) False
    (g) False
    (i) True

4. (a)

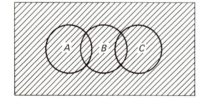

(c)

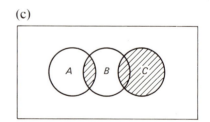

(e)

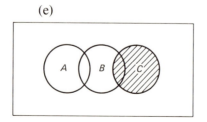

(g)

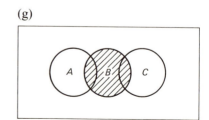

(i)

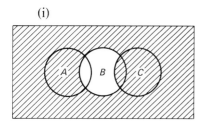

(k)

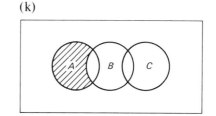

(m)

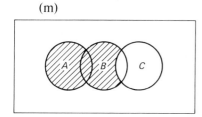

(o)

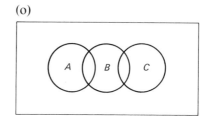

5. (a) $\varnothing$
    (c) $\varnothing$
    (e) A
    (g) $A \cup B$
    (i) A

Exercise Set 2.3
1. (a) $\{(0,1), (0,2), (0,3), (1,1), (1,2), (1,3), (2,1), (2,2), (2,3)\}$
    (c) $\{(0,3), (0,4), (0,5), (1,3), (1,4), (1,5), (2,3), (2,4), (2,5)\}$
    (e) $\{(0,3), (0,4), (0,5), (1,3), (1,4), (1,5), (2,3), (2,4), (2,5),$
         $(3,3), (3,4), (3,5)\}$
    (g) $\{(0,3), (0,4), (0,5), (1,3), (1,4), (1,5), (2,3), (2,4), (2,5),$
         $(3,3), (3,4), (3,5)\}$
    (i) $\{0, 1, 2, (1,3), (1,4), (1,5), (2,3), (2,4), (2,5), (3,3), (3,4),$
         $(3,5)\}$
    (k) $\{(0,0), (0,4), (0,5), (1,0), (1,4), (1,5), (2,0), (2,4), (2,5)\}$
    (m) $\{(1,0), (1,1), (1,2), (2,0), (2,1), (2,2)\}$
    (o) $\varnothing$

2. (a)   (c)

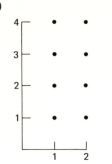   

(e)

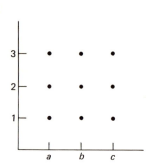

3. (a) ∅

Exercise Set 2.4

1. (a) Commutative property of union
    (c) Distributive property of Cartesian product over union
    (e) DeMorgan's laws
    (g) Distributive property of Cartesian product over intersection
    (i) Distributive property of union over intersection
2. (a) {(1,6), (2,6)} distributive property of Cartesian product over intersection
    (c) {$a,b,5$} distributive property of union over intersection
    (e) {0,1,2,3,4,5,7,8,9,10} DeMorgan's laws
    (g) {0,1,2,3} distributive property of union over intersection
    (i) ∅ distributive property of Cartesian product over intersection
5. (a) 8   (c) 3

Exercise Set 2.5

1. (a) Reflexive, transitive
   (c) Transitive
   (e) Reflexive, symmetric, transitive
   (g) Symmetric
   (i) None
2. (a) Transitive
   (c) None
3. (a) Is
   (c) Is
   (e) Is

Exercise Set 2.6

1. (a) {2,3,5}
   (b) {0,2,3,4,5}
   (c) {(0,3), (0,5), (2,3), (2,5), (4,3), (4,5)}
   (d) {(0,3), (0,5), (2,3), (2,5), (4,3), (4,5)}
   (e) ∅
   (f) ∅
   (g) {0, 2, 4, (1,2), (1,3), (1,5), (3,2), (3,3), (3,5), (5,2), (5,3), (5,5)}
   (h) {2,3,5}
   (i) {1,2,3,5}
   (j) {3,5}
   (k) {0,4}
   (l) ∅
   (m) {0,4}
   (n) {2,3,5}
   (o) ∅

2. (a)   (c)

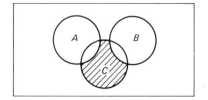

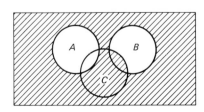

(e)

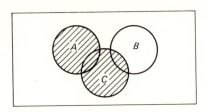

3. (a) $A \subset B$
   (c) $A = \varnothing$
   (e) $B \subset A$
4. (a) Transitive
   (c) Symmetric
   (e) Symmetric and transitive
   (g) Reflexive and transitive
   (i) Symmetric and transitive
5. (a) No   (e) No
   (c) No   (i) No
6. (a) No   (g) No
   (c) No   (i) No
   (e) No

Exercise Set 3.1
1. (a) 5   (i) 19
   (c) 9   (k) 6
   (e) 20   (m) 45
   (g) 6   (o) 5
2. (a) False   (g) False
   (c) False   (i) False
   (e) True
3. (a) Yes
   (c) $n(A \cap B)$
4. (a) 13
   (c) 4
5. (a) 25
   (c) 13
6. (a) 50
   (c) 5
   (e) 30

Exercise Set 3.2

1. (a) 720
   (c) 95
   (e) 12
   (g) $\frac{4}{3}$
   (i) $\frac{5}{16}$

2. (a) 567, 576, 657, 675, 765, 756
   (c) HT, TH
   (e) rag, rga, gra, gar, arg, agr

3. (a) 120
   (c) 720
   (e) 24
   (g) 3,628,800
   (i) 151,200
   (k) 210
   (m) 8,892,185,702,400
   (o) 1,860,480

4. (a) 2
   (c) 5040
   (e) 360
   (g) 45,360
   (i) 60

5. (a) $2^5 = 32$
   (c) $2^3 = 8$
   (e) $2^{10} = 1024$

6. (a) 6
   (c) 2
   (e) 12
   (g) 20
   (i) 20

7. (a) 2
   (c) $2^3 = 8$
   (e) $2^5 = 32$

9. 30,240

11. $10! = 3,628,800$

13. (a) 650 two-letter words
    (c) 358,800 four-letter words

Exercise Set 3.3

1. (a) *ab, ac, bc*
   (c) *rg, rb, ry, gb, gy, by*
   (e) *ab, ac, ad, ae, af, bc, bd, be, bf, cd, ce, cf, de, df, ef*

2. (a) 3
   (c) 35
   (e) 455

(g) 1
(i) 2
(k) 210
(m) 18,564
(o) 15,504

3. $\binom{n}{n-r} = \dfrac{n!}{[n-(n-r)]!(n-r)!} = \dfrac{n!}{r!(n-r)!} = \binom{n}{r}$

5. 495
7. 2,598,960
9. 20

Exercise Set 3.4

1. (a) $x^3 + 3x^2(-5b) + 3x(-5b)^2 + (-5b)^3 = x^3 - 15x^2b + 75xb^2 - 125b^3$

    (c) $(2x)^3 + 3(2x)^2(3) + 3(2x)(3)^2 + (3)^3 = 8x^3 + 36x^2 + 54x + 27$

    (e) $3^4 + 4(3^3)(2a) + 6(3)^2(2a)^2 + 4(3)(2a)^3 + (2a)^4 = 81 + 216a + 216a^2 + 96a^3 + 16a^4$

2. (a) $a^3 + 6a^2b + 12ab^2 + 8b^3$

    (c) $16a^4 + 96a^3b + 216a^2b^2 + 216ab^3 + 81b^4$

    (e) $32a^5 - 560a^4b + 3920a^3b^2 - 13{,}720a^2b^3 + 24{,}010ab^4 - 16{,}807b^5$

3. (a) 35
    (c) 35
    (e) 3432
    (g) 120
    (i) 8

4. (a) $3432x^7y^7$
    (c) $x^{14} + 42x^{13} + 819x^{12}$
    (e) $100{,}000 - 200{,}000 + 160{,}000$

Exercise Set 3.5

1. (a) Yes
    (c) No
    (e) No

2. (a) 10
    (c) 70
    (e) 28
    (g) 210
    (i) 1260
    (k) 2520
    (m) 126
    (o) 3,488,400

3. (a) 12
    (c) 504
    (e) 138,600
5. 1680
7. 13,860
9. 623,360,743,125,120

Exercise Set 3.6

1. (a) 10
    (c) 2
    (e) 0
    (g) 10
    (i) 11
2. (a) 6
    (c) 72
    (e) 3360
    (g) 863,040
    (i) 420
3. (a) 3
    (c) 36
    (e) 560
    (g) 35,960
    (i) 21
4. (a) 18,480
    (c) 120
    (e) 756,756
    (g) 126
    (i) 465,120
6. (a) $66a^{10}b^2$
    (c) 35
    (e) 220
    (g) 12,600
7. (a) False
    (c) True
    (e) False
9. 6 weeks
11. 240
12. (a) 36
    (c) 15
13. 1680, 60
15. (a) 126,126,000
    (b) Approximately 240 yr
17. 132

Exercise Set 4.1

1. (a) {1,2}
    (c) {1,2,3,4,5,6}
    (e) {HHH,HHT,HTH,HTT,THH,THT,TTH,TTT }
    (g) {1,2,3,4,5,6,7,8,9,10}

2. (a) $A = \{H1,H3,H5,T1,T2,T3,T4,T5,T6\}$
    (c) $\bar{C} = \{H1,H2,H3,H4,H5,H6,T2,T4,T6\}$
3. (a) $A \cup B = \{H2,H4,H6,H1,H3\}$
    (c) $A \cup C = \{H2,H4,H6,T1,T3,T5\}$
    (e) $A \cup D = \{H2,H4,H6,T5,T6\}$
    (g) $C \cap D = \{T5\}$
    (i) $B \cup D = \{H1,H2,H3,T5,T6\}$
    (k) $A \cap \bar{D} = \{H2,H4,H6\}$
    (m) $A \cap D = \{T5,T6\}$
    (o) $B \cap \bar{C} = \{H1,H2,H3\}$
    (q) $\bar{A} \cap C = \{T1,T3,T5\}$
    (s) $\bar{C} \cap \bar{D} = \{H1,H2,H3,H4,H5,H6,T2,T4\}$

Exercise Set 4.2

1. (a) 1/2
    (c) 3/8
2. (a) 1/4
    (c) 3/4
    (e) 7/8
3. (a) 1/4
    (c) 1/13
    (e) 1/2
    (g) 3/13
    (i) 3/26
4. (a) 7/15
    (c) 1/3
    (e) 1/15
    (g) 2/15
    (i) 13/15
5. (a) 1/6
    (c) 5/36
    (e) 1/36
    (g) 5/36
    (i) 1/6

    (e) 1/4
    (g) 1
    (g) 1/2
    (i) 1/2

    (k) 1/26
    (m) 4/13
    (o) 6/13
    (q) 9/13
    (s) 1/13
    (k) 2/5
    (m) 1/3
    (o) 1
    (q) 1/5
    (s) 1/15
    (k) 5/12
    (m) 1/18
    (o) 5/12
    (q) 1/6
    (s) 1/12

Exercise Set 4.3

1. (a) 1/6
    (c) $1 - (5/6) \cdot (5/6) \cdot (5/6) = 1 - (125/216) = 91/216$
    (e) $(1/6) \cdot (1/6) = 1/36$

    (g) $(1/2) \cdot (1/2) = 1/4$

    (i) $(1/6) \cdot (1/6) \cdot (1/6) \cdot (1/6) \cdot (1/6) \cdot (1/6) = (1/6)^6$

2. (a) $(3/7) \cdot (2/7) = 6/49$

    (c) $1 - (3/7) = 4/7$

    (e) $1 - (6/49) = 43/49$

    (g) $(4/7) \cdot (2/7) = 8/49$

    (i) $(3/7) \cdot (5/7) = 15/49$

3. (a) $(1/2) \cdot (1/3) \cdot (1/6) = 1/36$

    (c) $(1/2) \cdot (1/2) = 1/4$

    (e) $1 - (2/3) \cdot (2/3) = 1 - (4/9) = 5/9$

    (g) $[(1/2) \cdot (1/2)] + [(1/2) \cdot (1/2)] = 1/2$

    (i) $(1/2) \cdot (1/2) \cdot (1/2) = 1/8$

4. (a) $5/36$

    (c) $(1/2) \cdot (1/2) = 1/4$

    (e) $1 - (5/6) \cdot (5/6) = 1 - (25/36) = 11/36$

    (g) $(1/6) \cdot (5/6) + (5/6) \cdot (1/6) = (5/36) + (5/36) = 5/18$

    (i) $(5/6) \cdot (5/6) = 25/36$

5. **No**

Exercise Set 4.4

1. (a) $2/9$

    (c) $4/9$

    (e) $(3/10) \cdot (2/9) = 1/15$

2. (a) $1/4$ (g) $1/52$

    (c) $1/2$ (i) $1/13$

    (e) $1/13$ (k) $3/52$

3. (a) $(2/52) \cdot (1/51) = 1/1326$

    (c) $4/17$

    (e) $(26/52) \cdot (25/51) = 25/102$

    (g) $(13/52) \cdot (13/51) = 13/204$

    (i) $4/51$

    (k) $1/51$

    (m) $(26/52) \cdot (26/51) = 13/51$

    (o) $1/17$

4. (a) $(2/52) \cdot (2/52) = 1/676$

    (c) $1/4$

    (e) $1/4$

    (g) $1/16$

    (i) $1/13$

(k) 1/52

(m) 1/4

(o) 1/13

5. (a) 1/6   (g) $(1/6) \cdot (1/2) = 1/12$

(c) 1/2   (i) 3/5

(e) 1/6

6. (a) 3/17

(c) 1/4

(e) $(3/18) \cdot (2/17) \cdot (1/16) = 1/816$

(g) $(3/18) \cdot (2/17) \cdot (6/16) = 1/136$

(i) $1 - (12/18) \cdot (11/17) = 29/51$

(k) $(3/18) \cdot (15/17) + (15/18) \cdot (3/17) = 5/17$

(m) $(6/18) \cdot (3/17) \cdot (4/16) \cdot (5/15) = 1/204$

(o) $1 - (14/18) \cdot (13/17) \cdot (12/16) \cdot (11/15) = 2059/3060$

7. (a) $(1/4) \cdot (1/4) \cdot (1/4) = 1/64$

(c) $1 - (3/4) \cdot (3/4) \cdot (3/4) = 37/64$

(e) $(1/2) \cdot (1/2) \cdot (1/2) = 1/8$

(g) $3[(1/2) \cdot (1/2) \cdot (1/2)] = 3/8$

(i) $(1/2) \cdot (1/2) \cdot (1/2) + (1/4) \cdot (1/4) \cdot (1/4) + (1/4) \cdot (1/4) \cdot (1/4) = 5/32$

8. (a) $(25/100) \cdot (24/99) \cdot (23/98) = 23/1617$

(c) $1 - (75/100) \cdot (74/99) \cdot (73/98) = 3767/6468$

(e) $(50/100) \cdot (49/99) \cdot (48/98) = 4/33$

(g) $3[(50/100) \cdot (50/99) \cdot (49/98)] = 25/66$

(i) $(50/100) \cdot (49/99) \cdot (48/98) + (25/100) \cdot (24/99) \cdot (23/98) = 242/1617$

Exercise Set 4.5

1. (a) 23/70

(c) 18/35

(e) 1/35

(g) $2/7 + 17/35 - 1/7 = 22/35$

(i) 11/23

(k) 5/17

(m) 21/34

(o) 13/35

2. (a) 1/9

(c) 4/9

3. (a) 35/1000
    (c) 11/1000
    (e) 329/1000
    (g) 11/35
    (i) 6/35
    (k) 294/965
4. (a) 0.20
    (c) 0
    (e) 0.80

Exercise Set 4.6

1. (a) $\binom{20}{10}(3/5)^{10}(2/5)^{10}$

    (c) $\binom{20}{15}(3/5)^{15}(2/5)^{5} + \binom{20}{16}(3/5)^{16}(2/5)^{4} +$
         $\binom{20}{17}(3/5)^{17}(2/5)^{3} + \binom{20}{18}(3/5)^{18}(2/5)^{2} +$
         $\binom{20}{19}(3/5)^{19}(2/5) + (3/5)$

2. (a) $\binom{50}{10}(1/5)^{10}(4/5)^{40}$

    (c) $\binom{5}{1}(1/5)^{1}(4/5)^{4} + \binom{5}{2}(1/5)^{2}(4/5)^{3} + \binom{5}{3}(1/5)^{3}(4/5)^{2} +$
         $\binom{5}{4}(1/4)^{4}(4/5)^{1} + \binom{5}{5}(1/5)^{5}$

3. (a) 8/125
    (c) 2072/3125
    (e) 144/625
5. 5/16

Review Exercise Set 4.7

1. (a) 1/4
    (c) 1/2
    (e) 4/9
2. (a) 1/8
    (c) 7/8
    (e) 1/2
    (g) 1/4
    (i) 7/16

(k) 35/128

(m) 49/28,561

(o) 60/371,293

4. (a) 179/2000   (g) 49/179

(c) 1/200   (i) 1141/1821

(e) 83/200

Exercise Set 5.1

1. (a)

96–100	5		66– 70	5
91– 95	8		61– 65	6
86– 90	6		56–- 60	0
81– 85	8		51– 55	2
76– 80	8		46– 50	2
71– 75	5		41– 45	6

(b)

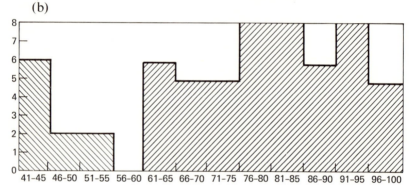

(c)

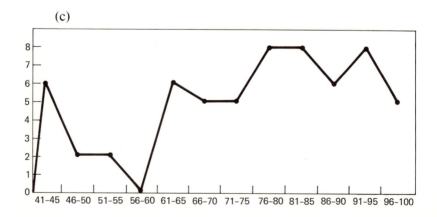

3. (a) A   3
         A−   2
         B+   4
         B   6
         B−   3
         C+   8
         C   14
         C−   8
         D+   2
         D   4
         D−   1
         F   3

(b)

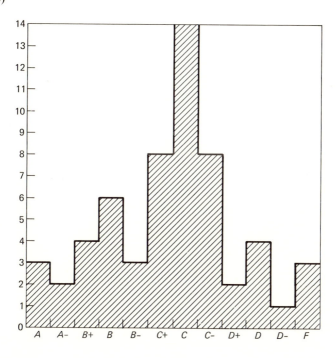

(c)

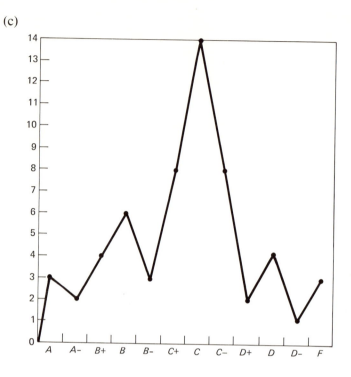

Exercise Set 5.2
1. (a) 12.8
   (b) 9
   (c) 8
3. (a) 23.3
   (b) 22
   (c) 22
5. $m = 76.3$
   $M = 76-80$
   Mode = 76–80
7. (Verify)

Exercise Set 5.3
1. (a) 64
   (b) 28
   (c) AD ≈ 5.7
   (d) V ≈ 7.9
   (e) SD ≈ 8.4

3. (a) 6'2"
    (b) 6'1"
    (c) 10"

    (d) AD ≈ 2.8"
    (e) SD ≈ 3.6"

Exercise Set 5.4
1. (a) 0.9987
    (c) 0.6554
    (e) 0.9993
    (g) 0.0179
2. (a) 0.4495
    (c) 0.0801
    (e) 0.0817
3. (a) 11/12
    (c) 4
    (e) −4/3
    (g) 45/12
4. (a) 0.0228
    (c) 0.1587
    (e) 0.0228
5. (a) 0
    (c) 0.6090
6. (a) 16
    (c) 1

    (i) 0.1841
    (k) 0.6985
    (m) 0.9633
    (o) 0.0040
    (g) 0.8904
    (i) 0.0056

    (i) 29/6
    (k) −11/6
    (m) 0
    (o) −1/4
    (g) 0.9759
    (i) 0.0215

Exercise Set 5.5
1. (a) ≈ 14,167
    (b) ≈ 3.3
    (c) ≈ 7470
3. (a) ≈ 3
    (b) ≈ 2.8
    (c) ≈ 2.2

    (d) ≈ 2.18
    (e) ≈ 0.72

    (d) ≈ .91
    (e) ≈ −0.96

Exercise Set 5.6
 1. 71
 3. 75
 5. 75
 7. ≈ 16.7
 9. ≈ 16
11. 50
13. −0.04

Exercise Set 6.1

1. (a) $\sqrt{26}$
   (b) $\sqrt{25} = 5$
   (c) $\sqrt{13}$
   (d) $\sqrt{37}$
   (e) 4
   (f) $\sqrt{21}$
   (g) $\sqrt{10}$
   (h) $\sqrt{14}$

   (i) $\sqrt{21}$
   (j) 5
   (k) $\sqrt{39}$
   (l) $\sqrt{21}$
   (m) $\sqrt{50} = 5\sqrt{2}$
   (n) $\sqrt{26}$
   (o) $\sqrt{164} = 2\sqrt{41}$

3. (a) True
   (b) True
   (c) True
   (d) False
   (e) True

   (f) True
   (g) True
   (h) True
   (i) False
   (j) True

Exercise Set 6.2

1. (a) $\begin{bmatrix} -2 & 5 \\ 3 & 1 \end{bmatrix}$

   (c) $\begin{bmatrix} -5 & 2 \\ 1 & -3 \\ 3 & 2 \end{bmatrix}$

   (e) $\begin{bmatrix} 7 & 4 & 9 \\ 11 & 8 & 6 \\ 13 & 17 & -3 \end{bmatrix}$

   (g) $\begin{bmatrix} \frac{2}{3} \\ 4 \\ \pi \end{bmatrix}$

   (i) $\begin{bmatrix} 7 & 11 & 2 & 4 \end{bmatrix}$

2. (a) $\begin{bmatrix} 2 & 1 & -1 \\ 2 & 1 & 13 \end{bmatrix}$

   (c) $\begin{bmatrix} 8 & 2 & -6 \\ 2 & 6 & 33 \end{bmatrix}$

   (e) $\begin{bmatrix} 24 & -2 & -30 \\ -26 & 26 & 33 \end{bmatrix}$

   (g) $\begin{bmatrix} -22 & 6 & 31 \\ 35 & -28 & 11 \end{bmatrix}$

(i) $\begin{bmatrix} 10 & -4 & -17 \\ -23 & 14 & -13 \end{bmatrix}$

(k) Undefined

(m) $\begin{bmatrix} 0 & 5 \\ 1 & -1 \\ 2 & 7 \end{bmatrix}$

(o) $\begin{bmatrix} 50 & 5 & -45 \\ -10 & 45 & 135 \end{bmatrix}$

3. (a) $\begin{bmatrix} -8 & 13 \\ -3 & 6 \end{bmatrix}$

(c) $\begin{bmatrix} 6 & -11 & 6 & -1 \\ -3 & -1 & -12 & -5 \\ 9 & 2 & -8 & -8 \end{bmatrix}$

(e) $\begin{bmatrix} -2 & -8 & 0 & -4 \\ 1 & 13 & -4 & 19 \\ -4 & 11 & -9 & -13 \end{bmatrix}$

Exercise Set 6.3

1. (a) 40

(c) $[51 \quad 18]$

(e) $\begin{bmatrix} 6 \\ 11 \end{bmatrix}$

(g) $\begin{bmatrix} 5 \\ -1 \\ 4 \end{bmatrix}$

(i) $\begin{vmatrix} 30 & -11 & -13 & 12 & 16 \\ 4 & 1 & -3 & 6 & 2 \\ 52 & -61 & -1 & -54 & 30 \end{vmatrix}$

(k) $|-26 \quad 4 \quad -28|$

(l) $\begin{vmatrix} 10 & 2 & 6 \\ 0 & 8 & 16 \\ 18 & -12 & 24 \end{vmatrix}$

3. 9×12

5. Let $\mathbf{A} = \begin{bmatrix} 0 & 0 \\ 1 & 0 \end{bmatrix}$ $\mathbf{B} = \begin{bmatrix} 0 & 1 \\ 0 & 0 \end{bmatrix}$

(a) $(\mathbf{A} + \mathbf{B})^2 = (\mathbf{A} + \mathbf{B})(\mathbf{A} + \mathbf{B})$

$$= \begin{bmatrix} 0 & 1 \\ 1 & 0 \end{bmatrix} \begin{bmatrix} 0 & 1 \\ 1 & 0 \end{bmatrix}$$

$$= \begin{bmatrix} 1 & 0 \\ 0 & 1 \end{bmatrix}$$

$$\mathbf{A}^2 + 2\mathbf{AB} + \mathbf{B}^2 = \begin{bmatrix} 0 & 0 \\ 1 & 0 \end{bmatrix} \begin{bmatrix} 0 & 0 \\ 1 & 0 \end{bmatrix} + 2 \begin{bmatrix} 0 & 0 \\ 1 & 0 \end{bmatrix} \begin{bmatrix} 0 & 1 \\ 0 & 0 \end{bmatrix}$$

$$+ \begin{bmatrix} 0 & 1 \\ 0 & 0 \end{bmatrix} \begin{bmatrix} 0 & 1 \\ 0 & 0 \end{bmatrix}$$

$$= \begin{bmatrix} 0 & 0 \\ 0 & 0 \end{bmatrix} + 2 \begin{bmatrix} 0 & 0 \\ 0 & 1 \end{bmatrix} + \begin{bmatrix} 0 & 0 \\ 0 & 0 \end{bmatrix}$$

$$= 0 + \begin{bmatrix} 0 & 0 \\ 0 & 2 \end{bmatrix} + 0$$

$$= \begin{bmatrix} 0 & 0 \\ 0 & 2 \end{bmatrix}.$$

$$\begin{bmatrix} 1 & 0 \\ 0 & 1 \end{bmatrix} \neq \begin{bmatrix} 0 & 0 \\ 0 & 2 \end{bmatrix}$$

$$\therefore (\mathbf{A} + \mathbf{B})^2 \neq \mathbf{A}^2 + 2\mathbf{AB} + \mathbf{B}^2.$$

(b) $(\mathbf{A} - \mathbf{B})(\mathbf{A} + \mathbf{B}) = \begin{bmatrix} 0 & -1 \\ 1 & 0 \end{bmatrix} \begin{bmatrix} 0 & 1 \\ 1 & 0 \end{bmatrix}$

$$= \begin{bmatrix} -1 & 0 \\ 0 & 1 \end{bmatrix}$$

$$\mathbf{A}^2 - \mathbf{B}^2 = \begin{bmatrix} 0 & 0 \\ 0 & 0 \end{bmatrix} - \begin{bmatrix} 0 & 0 \\ 0 & 0 \end{bmatrix}$$

$$= \begin{bmatrix} 0 & 0 \\ 0 & 0 \end{bmatrix}.$$

$$\begin{bmatrix} -1 & 0 \\ 0 & 1 \end{bmatrix} \neq \begin{bmatrix} 0 & 0 \\ 0 & 0 \end{bmatrix}$$

$$\therefore (\mathbf{A} - \mathbf{B})(\mathbf{A} + \mathbf{B}) \neq \mathbf{A}^2 - \mathbf{B}^2.$$

Exercise Set 6.4

1. cof. (3) $= -18$ cof. (1) $= -32$ cof. (4) $=$ 6
    cof. (2) $=$ 11   cof. (0) $=$ 23   cof. (6) $= -14$
    cof. (−5) $=$ 6   cof. (3) $= -10$ cof. (1) $= -2$

3. (a) (none)
    (b) −26
    (c) 30
    (d) 0
    (e) 70
5. (a) 62
    (c) 62
    (e) 0

(f) −8
(g) 12
(h) 65
(i) 309
(j) 8
(g) 0
(i) −3844
(k) 496

Exercise Set 6.5
1. (a) Nonsingular
    (c) Singular
    (e) Nonsingular
    (g) Nonsingular
    (i) Singular
3. (a) $\begin{bmatrix} 1 & 0 \\ 0 & 1 \end{bmatrix}$

    (c) $\dfrac{1}{31}\begin{bmatrix} 2 & -7 \\ 3 & 5 \end{bmatrix}$

    (e) Doesn't exist

    (g) $\dfrac{1}{65}\begin{bmatrix} -9 & 32 & -10 \\ -4 & 7 & 10 \\ 20 & -35 & 15 \end{bmatrix}$

    (i) $\dfrac{1}{20}\begin{bmatrix} 0 & 0 & 4 \\ 20 & -15 & -6 \\ 0 & 5 & -2 \end{bmatrix}$

Exercise Set 6.6
1. (a) $\begin{bmatrix} 1 & 0 \\ 0 & 1 \end{bmatrix}$

    (c) $\begin{bmatrix} 1 & 0 \\ 0 & 1 \end{bmatrix}$

    (e) $\begin{bmatrix} 1 & 0 \\ 0 & 1 \\ 0 & 0 \end{bmatrix}$

(g) $\begin{bmatrix} 1 & 0 & 0 \\ 0 & 1 & 0 \\ 0 & 0 & 1 \end{bmatrix}$

(i) $\begin{bmatrix} 1 & 0 & 0 & \dfrac{106}{33} \\ 0 & 1 & 0 & \dfrac{157}{33} \\ 0 & 0 & 1 & \dfrac{17}{3} \end{bmatrix}$

Exercise Set 6.7

1. $x = 3$, $y = 4$
3. $x = y = 1$
5. $x = 3/2$, $y = 7/4$
7. $x = 1$, $y = -4$
9. $x = 4/5$, $y = 11/10$
11. $x = -2$, $y = 2$
13. $x = 1$, $y = -3$, $z = 2$
15. $x = 1$, $y = 2$, $z = -3$
19. $x = y = z = 1$

Exercise Set 6.8

1. (a) $(9,-4,4)$
   (b) $(1,2,0)$
   (c) $(22,-11,10)$
   (d) 27
   (e) 27
   (f) $\sqrt{30}$
   (g) $\sqrt{29}$
   (h) $\sqrt{113}$
   (i) $\sqrt{30} + \sqrt{29}$
   (j) 1
3. (a) $x = 0$, $y = 3$
   (b) $x = 3$, $y = 2$
   (c) $x = 0$, $y = 3$
   (d) No solution
   (e) $x = 1$, $y = -1$
   (f) Infinitely many solutions

(g) $x = 2$, $y = 1$, $z = 3$

(h) $x = z = 1$, $y = 3$

(i) $x = -59/21$, $y = 58/21$, $z = 23/21$

(j) $x = 3$, $y = 2$, $z = 4$

(k) $x = 0$, $y = 2$, $z = 0$

(l) No solution

Exercise Set 7.1

1. (a)

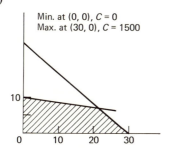

Min. at (0, 0), $C = 0$
Max. at (30, 0), $C = 1500$

(c)

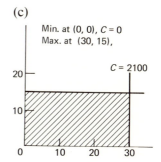

Min. at (0, 0), $C = 0$
Max. at (30, 15),

$C = 2100$

(e)

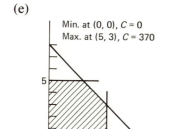

Min. at (0, 0), $C = 0$
Max. at (5, 3), $C = 370$

(g)

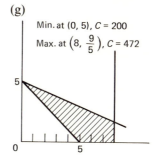

Min. at (0, 5), $C = 200$

Max. at $\left(8, \frac{9}{5}\right)$, $C = 472$

(i)

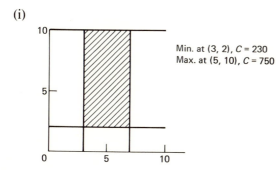

Min. at (3, 2), $C = 230$
Max. at (5, 10), $C = 750$

3. (a) 128 packages of A and 32 packages of B. Income = $89.60
   (c) 140 packages of A. Income = $140.00
4. (a) 23 homes of type A and 30 homes of type B. Income = $945,000
   (c) 39 homes of type A and 17 homes of type B. Income = $1,400,000
   (e) 39 homes of type A and 17 homes of type B. Income = $1,035,000
5. (a) 240 sweaters of type A and 240 sweaters of type B, or 480 sweaters of type A and 0 sweaters of type B, or any other combination of 480 sweaters as long as at least 240 are of type A
   (c) No sweaters of type A and 400 sweaters of type B
7. He should raise seed on all 200 acres. His maximum profit is $60,000.

Exercise Set 7.2

3. (a) $x_1 = 0$, $x_2 = 7$, $f_{max} = 7$
   (c) $x_1 = \frac{9}{5}$, $x_2 = \frac{11}{10}$, $x_3 = 0$, $f_{max} = 13\frac{2}{5}$
   (e) $x_1 = 2$, $x_2 = 1$, $f_{min} = 16$
5. 20 pairs of pants, $12\frac{1}{2}$ sport coats, and $12\frac{1}{2}$ coats. Maximum profit = $950
7. 6 sacks of wheat and 4 sacks of rye

Exercise Set 7.3

1. (a) -8
   (c) 0
   (e) $-5/3$
2. (a) Strictly determined. -4 is the saddle point. Optimal strategies are $\mathbf{P} = [0,1]$ and

$$\mathbf{Q} = \begin{bmatrix} 1 \\ 0 \end{bmatrix}.$$

   (c) Not strictly determined
   (e) Not strictly determined

3. (a)   H   T

$$\begin{array}{c} H \\ T \end{array} \begin{bmatrix} 1 & -1 \\ -1 & 1 \end{bmatrix}$$

(b)   1   2   3

$$\begin{array}{c} 1 \\ 2 \\ 3 \end{array} \begin{bmatrix} 2 & -2 & 4 \\ -2 & 4 & -6 \\ 4 & -6 & 6 \end{bmatrix}$$

Exercise Set 7.4

1. (a) Strictly determined. 3 is the saddle point. The optimal strategies are $\mathbf{P}' = [0,1]$ and

$$\mathbf{Q}' = \begin{bmatrix} 0 \\ 1 \end{bmatrix}.$$

(c) Strictly determined. 0 is the saddle point. The optimal strategies are $\mathbf{P}' = [1,0]$ and

$$\mathbf{Q}' = \begin{bmatrix} 1 \\ 0 \end{bmatrix}.$$

(e) Nonstrictly determined. Optimal strategies are

$$\left[\frac{-4-2}{-9-12}, \frac{-5-10}{-9-12}\right] = \left[\frac{2}{7}, \frac{5}{7}\right]$$

$$\mathbf{Q}' = \begin{bmatrix} \dfrac{-4-10}{-9-12} \\ \dfrac{-5-2}{-9-12} \end{bmatrix} = \begin{bmatrix} \dfrac{2}{3} \\ \dfrac{1}{3} \end{bmatrix}.$$

(g) Nonstrictly determined. Optimal strategies are

$$\mathbf{P}' = \left[\frac{-6-4}{-8-7}, \frac{-2-3}{-8-7}\right] = \left[\frac{2}{3}, \frac{1}{3}\right]$$

$$\mathbf{Q}' = \begin{bmatrix} \dfrac{-6-3}{-8-7} \\ \dfrac{-2-4}{-8-7} \end{bmatrix} = \begin{bmatrix} \dfrac{3}{5} \\ \dfrac{2}{5} \end{bmatrix}.$$

(i) Nonstrictly determined. Optimal strategies are

$$\mathbf{P}' = \left[\frac{3-4}{1-9}, \frac{-2-5}{1-9}\right] = \left[\frac{1}{8}, \frac{7}{8}\right]$$

$$\mathbf{Q}' = \begin{bmatrix} \dfrac{3-5}{1-9} \\[2mm] \dfrac{-2-4}{1-9} \end{bmatrix} = \begin{bmatrix} \dfrac{1}{4} \\[2mm] \dfrac{3}{4} \end{bmatrix}.$$

(k) Nonstrictly determined. Optimal strategies are

$$\mathbf{P}' = \left[\frac{5-(-2)}{9-(-12)}, \frac{4-(-10)}{9-(-12)}\right] = \left[\frac{1}{3}, \frac{2}{3}\right]$$

$$\mathbf{Q}' = \begin{bmatrix} \dfrac{5-(-10)}{9-(-12)} \\[2mm] \dfrac{4-(-2)}{9-(-12)} \end{bmatrix} = \begin{bmatrix} \dfrac{5}{7} \\[2mm] \dfrac{2}{7} \end{bmatrix}.$$

2. (a) 3
   (c) 0
   (e) 0
   (g) 0
   (i) 13/4
   (k) 0
3. (a) Favors R
   (c) Fair
   (e) Fair
   (g) Fair
   (i) Favors R
   (k) Fair
4. (a)
$$\begin{array}{c c c} & H & T \\ H & 1 & -1 \\ T & -1 & 1 \end{array}$$

   Nonstrictly determined. Optimal strategies are $\mathbf{P} = [1/2, 1/2]$
   and

$$\mathbf{Q} = \begin{bmatrix} 1/2 \\ 1/2 \end{bmatrix}.$$

   $v = 0$. Thus the game is fair.
   (c)
$$\begin{array}{c c c} & R & P \\ R & 0 & 5 \\ P & 8 & 0 \end{array}$$

Nonstrictly determined. Optimal strategies are $\mathbf{P} = [8/13, 5/13]$ and

$$\mathbf{Q} = \begin{bmatrix} 5/13 \\ 8/13 \end{bmatrix}.$$

$v = \dfrac{40}{13}$. So the game favors the offense.

Exercise Set 7.5

1. (a) $\begin{bmatrix} 1 & 0 \\ 0 & 3 \end{bmatrix}$ (g) $\begin{bmatrix} -6 & 4 \\ 5 & 3 \end{bmatrix}$

   (c) $\begin{bmatrix} -5 & -3 \\ 5 & 3 \end{bmatrix}$ (i) $[-3]$

   (e) $\begin{bmatrix} -5 & 4 \\ 4 & 1 \end{bmatrix}$

2. (a) $\mathbf{P}' = [3/4, 1/4]$, $\mathbf{Q}' = \begin{bmatrix} 3/4 \\ 1/4 \end{bmatrix}$, $v = 3/4$

   (c) $\mathbf{P}' = [0,1]$, $\mathbf{Q}' = \begin{bmatrix} 0 \\ 1 \end{bmatrix}$, $v = 3$

   (e) $\mathbf{P}' = [1/4, 3/4]$, $\mathbf{Q}' = \begin{bmatrix} 1/4 \\ 3/4 \end{bmatrix}$, $v = 7/4$

   (g) $\mathbf{P}' = [1/6, 5/6]$, $\mathbf{Q}' = \begin{bmatrix} 1/12 \\ 11/12 \end{bmatrix}$, $v = 19/6$

   (i) $\mathbf{P}' = [1,0]$, $\mathbf{Q}' = \begin{bmatrix} 0 \\ 1 \end{bmatrix}$, $v = -3$

3. (a) Unfair
   (c) Unfair
   (e) Unfair
   (g) Unfair
   (i) Unfair
4. (a) No saddle point
   (c) 3
   (e) No saddle point
   (g) No saddle point
   (i) -3

5. $\mathbf{P}' = [1/2,1/2,0]$

6.   (a) $\mathbf{P}' = [7/8,1/8]$, $\mathbf{Q}' = \begin{bmatrix} 1/4 \\ 3/4 \end{bmatrix}$, $v = 17/4$

     (c) $\mathbf{P}' = [4/5,0,1/5]$, $\mathbf{Q}' = \begin{bmatrix} 3/5 \\ 2/5 \end{bmatrix}$, $v = 17/5 - 4 = -3/5$

     (e) $\mathbf{P}' = [0,1,0]$, $\mathbf{Q}' = \begin{bmatrix} 1 \\ 0 \end{bmatrix}$, $v = 3$

     (g) 0 is a saddle point, so $\mathbf{P}' = [0,1,0]$, $\mathbf{Q}' = \begin{bmatrix} 0 \\ 0 \\ 1 \end{bmatrix}$, and $v = 0$

     (i) $\mathbf{P}' = [1/4,3/4,0]$, $\mathbf{Q}' = \begin{bmatrix} 1/4 \\ 3/4 \end{bmatrix}$, $v = 31/4 - 8 = -1/4$

7.   (a) $\begin{bmatrix} 1 & -1 \\ -1 & 1 \end{bmatrix}$, $\mathbf{P}' = [1/2,1/2]$, $\mathbf{Q}' = \begin{bmatrix} 1/2 \\ 1/2 \end{bmatrix}$, $v = 0$

     (c) $\begin{bmatrix} 100 & 0 & -30 \\ 80 & 20 & -10 \\ 50 & 40 & 30 \end{bmatrix}$, $\mathbf{P}' = [0,0,1]$, $\mathbf{Q}' = \begin{bmatrix} 0 \\ 0 \\ 1 \end{bmatrix}$,

$$v = 30$$

     (e)   cool   cord.   char.
$$\begin{matrix} \text{cool} \\ \text{cord.} \\ \text{char.} \end{matrix} \begin{bmatrix} 0 & -1 & 2 \\ 1 & 0 & -1 \\ -2 & 1 & 0 \end{bmatrix}, \quad \mathbf{P}' = [1/4,1/2,1/4], \quad \mathbf{Q}' = \begin{bmatrix} 1/4 \\ 1/2 \\ 1/4 \end{bmatrix}$$

$$v = 0$$

INDEX